T0237565

Springer Theses

Recognizing Outstanding Ph.D. Research

Aims and Scope

The series "Springer Theses" brings together a selection of the very best Ph.D. theses from around the world and across the physical sciences. Nominated and endorsed by two recognized specialists, each published volume has been selected for its scientific excellence and the high impact of its contents for the pertinent field of research. For greater accessibility to non-specialists, the published versions include an extended introduction, as well as a foreword by the student's supervisor explaining the special relevance of the work for the field. As a whole, the series will provide a valuable resource both for newcomers to the research fields described, and for other scientists seeking detailed background information on special questions. Finally, it provides an accredited documentation of the valuable contributions made by today's younger generation of scientists.

Theses may be nominated for publication in this series by heads of department at internationally leading universities or institutes and should fulfill all of the following criteria

- They must be written in good English.
- The topic should fall within the confines of Chemistry, Physics, Earth Sciences, Engineering and related interdisciplinary fields such as Materials, Nanoscience, Chemical Engineering, Complex Systems and Biophysics.
- The work reported in the thesis must represent a significant scientific advance.
- If the thesis includes previously published material, permission to reproduce this must be gained from the respective copyright holder (a maximum 30% of the thesis should be a verbatim reproduction from the author's previous publications).
- They must have been examined and passed during the 12 months prior to nomination.
- Each thesis should include a foreword by the supervisor outlining the significance of its content.
- The theses should have a clearly defined structure including an introduction accessible to new PhD students and scientists not expert in the relevant field.

Indexed by zbMATH.

More information about this series at http://www.springer.com/series/8790

Feng Bao

Computational Reconstruction of Missing Data in Biological Research

Doctoral Thesis accepted by
Tsinghua University, Beijing, China

Author
Dr. Feng Bao (iD)
Department of Automation
Tsinghua University
Beijing, China

Supervisors
Prof. Qionghai Dai
Department of Automation
Tsinghua University
Haidian, Beijing, China

Prof. Jinli Suo
Department of Automation
Tsinghua University
Haidian, Beijing, China

ISSN 2190-5053   ISSN 2190-5061 (electronic)
Springer Theses
ISBN 978-981-16-3063-7   ISBN 978-981-16-3064-4 (eBook)
https://doi.org/10.1007/978-981-16-3064-4

Jointly published with Tsinghua University Press
The print edition is not for sale in China (Mainland). Customers from China (Mainland) please order the print book from: Tsinghua University Press.

This Springer imprint is published by the registered company Springer Nature Singapore Pte Ltd.
The registered company address is: 152 Beach Road, #21-01/04 Gateway East, Singapore 189721, Singapore

To my beloved grandparents, parents and wife, for the support over the years.

Supervisors' Foreword

It is our pleasure to introduce the Ph.D. Thesis authored by Dr. Feng Bao. This thesis was nominated by the Graduate School of Tsinghua University to publish with Springer Theses.

Dr. Bao joined the Institute of Brain and Cognitive Sciences, Department of Automation, Tsinghua University in 2014, where he received the Ph.D. of Control Science and Engineering degree in June 2019.

During Dr. Bao's Ph.D. study, his research focuses on machine learning and computational biology. He designed advanced statical algorithms and interrogated the complex systems of biological activities. The research in this thesis was motivated by the fact that currently biological studies produced large-scale—yet with huge missing entries—data. Dr. Bao categorized the missing problems in biological research into three major types and proposed computational recovery methods to accurately achieve multiple biological tasks, including cell dissection, brain segmentation, and function prediction. The advances of these new artificial intelligence methods enable the efficient and effective usage of the corrupted data and make the comprehensive understanding of complicated systems possible, especially on the brain. In particular, the thesis includes the following:

- Deep recurrent neural network for feature missing problem.
- Robust information theoretic learning for label missing problem.
- Structure sensing and rebalancing for sample missing problem.

This Ph.D. thesis was selected as an Excellent Ph.D. thesis of Tsinghua University. Dr. Bao published a number of works on Nature Methods, Patterns, Nucleic Acids Research, IEEE Trans. Neural Network and Learning Systems et al. His works provide basic tools to facilitate the biological data interpretations, and the power of proposed methods was not restricted to biology data but can be extended to other research fields. We believe the contribution of this thesis can be useful and appealing to the scientific community.

Beijing, China   Prof. Qionghai Dai
April 2021   Prof. Jinli Suo

Preface

The development of new life science technologies has significantly advanced the study of biological mechanisms, extending the boundaries of our understanding to life functionalities. With new advances, we are able to observe bioactivities at higher dimensions, larger scales, and deeper throughput. However, due to the immaturity of new technologies and the restriction in biological experiments, the collected biological data generally include lots of missing entries, e.g., missing features, missing labels and missing data samples. Data missing problems limit existing computational approaches to effectively reveal new discoveries from the data. This book introduces statistical recovery methods for large-scale and diverse biological genomic data and imaging data and provides effective tools for life science research from low-quality observation data. The main contributions and innovations are summarized as follows:

(1) Deep recurrent neural network recovery for feature missings. For missing problems of biological features, we propose a deep recurrent neural network recovery framework to learn the natural intrinsic correlation structure between the feature dimensions of biological data and reconstruct the missing information layer-by-layer through the designed recurrent network. The method has low computational consumption, low memory requirement and high reconstruction accuracy, which makes rapid large-scale data feature recovery possible. The computational framework is applied to analyze brain single-cell sequencing data in million scale.

(2) Robust information theoretic learning for label missings. The classification problem of biological data heavily relies on the experts' manual annotations. To relieve the requirement of experts' labeling, we propose the robust information theory learning method. Using the existing partial-labeled information, we can extract effective features and construct sample annotations under the information theory metrics. This computational method is applied to the partially labeled brain fMRI image segmentation task and identifies different brain structure regions with high precision.

(3) Structure-aware rebalancing for minor sample missings. For statistical analysis with rare samples, we propose structure-aware imbalance data reconstruction to improve the association analysis and label prediction tasks of existing

statistical models on imbalanced data. The methods are applied for the effective identification of high-risk pathogenic gene loci and prediction of rare cell subpopulations.

San Francisco, US   Feng Bao
August 2020

Acknowledgements

I sincerely thank Prof. Qionghai Dai and Associate Professor Jinli Suo for their supervision during my Ph.D. study. Dr. Dai showed me key properties to be a good scientist in academic research and encouraged me to explore interesting fields and keep a forward-looking insight. These will become my principle in my futural research career. During my Ph.D. study, Dr. Suo gave me detailed guidance and helped me build research directions and research methods. The experiences I learned in my Ph.D. years will benefit me for life. Thanks to Dr. Deng Yue's practical guidance in solving scientific problems.

Thanks to all teachers and students in Broadband Network Digital Media Lab, Tsinghua University, for their help and support. Special thanks to Ms. Rong Fang, Dr. Tao Yue, Dr. Liheng Bian, Dr. Xuemei Hu, Mr. Dongsheng An, Dr. You Zhou, Dr. Jiamin Wu, Dr. Ziwei Li, and Mr. Bo Xiong. I will always remember the days of working and studying with them.

Thanks to Prof. Guo-Cheng Yuan at Harvard University, Prof. Long Cai at California Institute of Technology, and Profs. Steven Altschuler and Lani Wu at University of California, San Francisco, for cooperation and discussion during my visiting study. Thanks to Dr. Mulong Du and Dr. Shengbao Suo for their care and help during the visiting study.

Thanks to my family for their unconditional trust and support for the research path I chose. Your continuing encouragement to me has helped me go further.

Thanks to the Editor of this book, Ms. Qian Wang, for the assistance in publication.

Thanks to the National Natural Science Foundation of China for supporting this project and Tsinghua University Doctoral Short-Term Visiting Fund for supporting the exchange study.

Contents

Acronyms

ARI	Adjusted Rand index
Bosco	Boosting correction method
DELTA	Deep landmark in latent space
DRIT	Deep robust information theoretic embedding
GWAS	Genome-wide association study
LASSO	Least absolute shrinkage and selection operator
MNN	Mutual nearest neighbors
ReLU	Rectified linear unit
RIT	Robust information theoretic embedding
scRNA-seq	Single-cell RNA sequencing
SS-RIT	Structural sparse robust information theoretic embedding
WRL	Weighted logistic regression
ZINB	Zero-inflated non-negative binomial distribution

Chapter 1
Introduction

Life science is an important research field that explores the basic mechanism of life and advances the understanding of biological activities. In the past decade, the vigorous development of new biological technologies has provided effective tools for life science study, making it possible to collect biological data and reveal the life science functionalities on large scale, deep level, and multiple angles. Deriving meaningful biological conclusions from biological data is the ultimate goal in life science research. However, due to the immaturity of new observation technology and the limitations of research problems, the collected data are often degraded, which restricts our analysis and processing . This chapter firstly reviews the major progress made in the field of life science observations in recent years, and the data missing problems that still exist in new technologies, and then discuss the existing computational solutions to reduce effects from data missings.

1.1 Research Background

In the research of life sciences, the advance of observation technology is an important driving force for the continuous development of the entire life science field. Many life science discoveries of great significance rely on the breakthrough of new observation technology. After years of development, these important biotechnologies have expanded from the field of basic life science research to the field of medical applications, becoming important tools for assisting the diagnosis and treatment of human diseases (Fig. 1.1).

For example, the computed tomography technology [1] (CT), which was awarded the Nobel Prize in Physiology in 1979, is an important way of high spatial resolution non-invasive medical 3D imaging. It makes use of differences in the absorption capacity of different tissues for X-rays and three-dimensional technology to reconstruct tomographic images, and are widely used in the medical diagnosis of brain and

© Tsinghua University Press 2021
F. Bao, *Computational Reconstruction of Missing Data in Biological Research*,
Springer Theses, https://doi.org/10.1007/978-981-16-3064-4_1

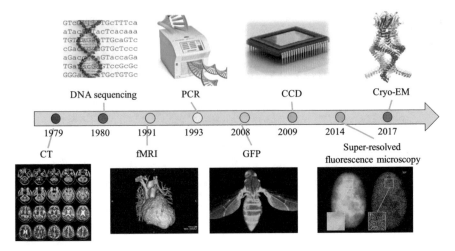

Fig. 1.1 Observation technologies in the field of life science have been awarded Nobel Prizes many times in history

heart diseases. DNA sequencing technology won the Nobel Prize in Chemistry in 1980 [2, 3]. Trough fluorescent markers of DNA bases, this technology enables precise measurement of human genetic basis and obtains information on human genetic mutations. They have been widely used in the diagnosis of human genetic diseases and the screening of therapeutic drug targets. High-resolution nuclear magnetic field won the Nobel Prize in Chemistry in 1991 [4, 5]. It uses nuclear magnetic resonance phenomenon to obtain the distribution information of hydrogen in the human body, thereby achieving accurate reconstruction of the internal structure. It has been a non-intrusive, non-radiative, fast and accurate clinical detection method that plays a key role in the diagnosis of many important diseases, such as cancer and Parkinson's disease. The gene amplification technology won the Nobel Prize in Chemistry in 1993 [6, 7]. It achieves a low-cost amplication approach of DNA or RNA fragments and makes it possible to detect genetic sequences with low expression levels. This technology provides an important way to understand the life activity of tissues and organs with high precision, and plays an important role in tumor detection and other fields. These technologies have been well-tested through the years, mark important contributions to the advances of life sciences and medical research and improve human health.

In recent years, life science observation technology has accelerated its development towards high precision, deep level, and wide field of vision. On an unprecedented scale, it has shown us the way how life functions. For example, the super-resolution fluorescence microscopy technology, which won the 2014 Nobel Prize in Chemistry [8, 9], uses the random activation characteristics of fluorescent proteins with optical computational reconstructions, to overcome the physical optical diffraction limitations. It enables the functional study of life activities at the subcellular precision and organization level. Nowadays it has played an important role in the research of tumor metastasis, brain neural connection, tissue structure and other fields. The cryo-freezing electron microscope [10, 11], which won the Nobel Prize in Chemistry in 2017, provides the three-dimensional spatial structure of the sample and has led to a major leap forward in the development of structural biology. On the basis of close to atomic resolution, it has deeply revealed the molecular structure of many important proteins, viruses, ribosomes and enzyme complexes.

With the continuous progress of research, new observation technologies are still emerging one after another, such as photoacoustic imaging [12, 13], which uses light to conduct thermal energy and converts position information into acoustic signals for reception, providing a non-invasive observation approach. Serialized single-molecule fluorescent labeling technology (seqFISH) and multiplexing, error-robust single-molecule fluorescent labeling technology [14] (merFISH), can directly recognize tens of thousands of RNA transcripts in a single cell while retaining their spatial localizations. This technology uses fluorescent probes to label sequences in multiple rounds, with each sequence has a unique corresponding fluorescent barcode to uniquely identify the RNA present in the sample, revealing the structural characteristics of chromosomes and the expression characteristics of cell activity at the single cell level. Multiplexing indirect immunofluorescence imaging technology [15] (4i), directly labels multiple proteins by multiplex immunofluorescence technology to achieve in situ abundance measurement of multiplexed proteins. Single-cell sequencing [16, 17] is a rapidly developing method for characterizing the genome at individual cells. The development of single cell technology and computing methods has made it possible to systematically analyze cell heterogeneity in a wide range of tissues and cell populations, and to explore the state composition, dynamic changes and regulatory mechanisms of cells in development and disease. These technologies provide a powerful tool for studying life science mechanisms from multiple perspectives on the biological samples.

1.2 Computational Reconstruction of Data Missings in Biological Study

The development of new technologies makes it possible to gain a deeper understanding of life sciences, and provides an important way to solve many basic biological problems [18], such as analyzing the cellular functions of tissues and organs, tracking embryonic development and so on. However, to draw effective life science conclusions from biological observation data, it requires a comprehensive data analysis matching the data properties of different technologies (Fig. 1.2). Statistical analysis of biological data is an important direction. Data preprocessing pipeline includes the following steps: (1) converting the original biological measurement data into unbiased signals with actual biological meanings; (2) the error correction of experimental batches; (3) missing data imputation; (4) data feature compression that effectively improves the quality of data analysis and enhance the exclusive characteristics of the

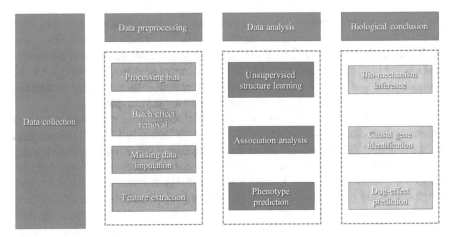

Fig. 1.2 Data analysis pipeline from raw biological data to biological conclusions

data; (5) direct data analysis using statistics, machine learning models on the processed data; (6) statistical analysis for specific tasks, such as data internal structure perception (clustering analysis), causality test of measured feature data and phenotypic data, data outcome prediction (classification prediction), etc.; (7)interpreting analysis results as biological conclusions, such as from clusters to the mechanism of life, from causal signals of disease to medicine targets of drugs.

In the data processing, an important bottleneck is how to perform statistical analysis on missing data. Data missing is a critical challenge in all types of biological data (Fig. 1.3), especially in applications of new technologies. As new approches are in the early stage of technology development, the experimental protocol of the technology is immature and is constrained by other current technology-dependent factors (e.g. the sensitivity of the optical detector, the efficiency of cell separation, etc.). As a result, many of data features are missing, e.g. many of the measured genes are lost, the brightness of the fluorescent marker is lower than the detectable range, etc. These lost features may contain important information about the sample properties, hindering the effective processing and exploration of the data. In addition, in real applications to many specific biomedical problems, there can be another two types of missing data: label missings and samples missings. To provide sample labels, most biological studies require professional inspections by biologists or medical scientists. Each sample (cell or individual) needs to be manually annotated based on professional experiences of experts and this annotation process is time-consuming. However, the life science data analysis relies heavily on large-scale data collection. Reliable conclusions can only be obtained if the sample scale is large enough. However, to label such a great number of samples, manual annotations consumes a lot of manpower and material resources, and becomes a key constraint. Therefore, in many real large-scale data sets, only a small number of samples have accurate labels. How to derive effective biological conclusions from partially labeled data poses chal-

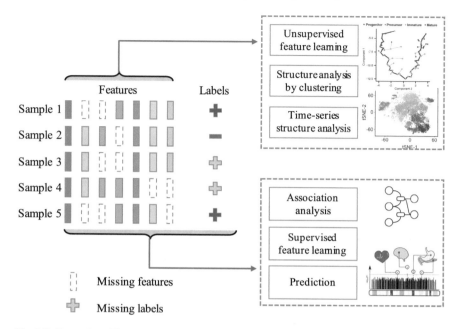

Fig. 1.3 Examples of feature missings and labels missings in biological data and types of statistical analyses on missing data

lenges to traditional computational methods. When the research involves some rare diseases, it is hard to collect a dataset of enough samples with certain phenotypes. In this case, analyses on a dataset with biased sample distributions will lead to conventional computational methods performing poorly. Thus, these data missing problems seriously limit the full exploration and effective usage of valuable biological data.

1.2.1 Feature Missings in Biological Data

The most typical case of missing feature information is in the process of single-cell RNA sequencing (scRNA-seq). scRNA-seq can quantify intrisic population heterogeneity at a high resolution and reveal dynamics of cell development and complex tissues variety. Recent advances in high-throughput sequencing technologies, including Dropseq [19], Indrop [20], provide larger scale data of cell profiles and are expected to solve various challenges in biology. The key bottleneck of applying scRNA-seq technology is to accurately measure the abundance of RNA expression in each cell. However, in the sequencing process, due to the limited ability of cell dissociation and sequence amplication, often only a small amount of RNA expression (10% ~ 30%) can be detected. Moreover, in the sequencing data involving large-scale of cells (millions to tens of millions), because the overall sequencing capacity has not been

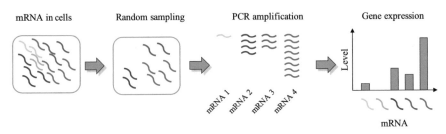

Fig. 1.4 Schematic diagram of the data processing pipeline in RNA sequencing and the source of missings

significantly improved, the average number of genes that can be collected on each cell will further decline ($\sim 5\%$). This low average capacity will leave many genes with empty measurements, especially those with low expression levels (Fig.1.4). In addition, there is inherent measurement noise in sequencing, which further increases the error of gene expressions. When performing cluster analysis on data that contains missings and noise, the true internal structure cannot be accurately identified and real cell types can be incorrectly discovered. Another typical case is the lack of observed fluorescent markers in fluorescence microscopic imaging. Especially in the new fluorescent imaging with multiplexed barcoding (seqFISH/merFISH) [14], where multiple rounds of fluorescence labeling, washing and bleaching are used at sequentially, it will lead to the fluorescence labeling efficiency decreasing gradually. At the same time, in the analysis of images from each imaging round, constrasts of different fluorescent channels vary greatly, and many darker markers cannot be distinguished from random background noise. Therefore, in the actually preprocessing pipeline on the data, it is often necessary to select a threshold to filter the obvious fluorescent markers. In seqFISH, only 20% of fluorescent labels can be accurately recorded. With such partial observations in the imaging data, complete biological structure information cannot be retained with a high resolution.

These missing data are the direct loss of the features from the observed sample itself, which has a great impact on the direct understanding and analysis of the information contained in the sample. Under the presence of current observation technology, it is valuable to use the computational method to handle the missing features and improve the sample observation results. To solve this problem, existing computational frameworks for data with missing features can be classified into two categories: feature imputations based on probability estimation and feature imputations based on computational reconstruction. For the first type of methods for imputation, statistical modeling of the feature generation process is performed first. In this step, probability distribution function of the feature is designed, and then the parameters in the probability distribution function are fitted and solved based on observed feature data. Finally, on the fitted probability distribution, the missing data is randomly sampled from the distribution for imputation. For example, in the single-cell analysis, the zero-inflatened non-negative binomial distribution (ZINB) is used to model the ramdonly gene count missing during the sequencing process. However, in statistical modeling, because of the actual data generation process is often complex and involves different distribution assumptions, the solution to model fitting can only be obtained through generalized methods such as Markov chain Monte Carlo (MCMC), to perform inference on the probabilistic graph. This method can be particularly inefficient when dealing with large-scale data. The other method is a direct reconstruction method based on the missing entries. In such methods, the reconstruction of missing data is the learning target function and a general model without specific probability assumptions is constructed. With model optimization on the existing observation data, parameters that achieve the minimum reconstruction error are obtained. With learned model, missing features can be imputed by directly running the generative process.

1.2.2 Label Missings in Biological Data

The lack of label information can be widely witnessed in many life science problems. Statistical analysis on partially labeled data is called semi-supervised learning in the field of machine learning. When data annotation is difficult or expensive, semi-supervised learning can employ readily available unlabeled data to improve supervised learning tasks. In real application, semi-supervised learning has great practical value. In many tasks, data tagging requires manual annotations, special equipments, or expensive and slow experiments. This problem is widely existed, e.g. in image classification, the content of each image needs to be labeled to provide a large-scale training set. However, for large-scale data containing millions of images, such as ImageNet, labeling requires a lot of manpower. In general machine learning problems, most of the data labeling does not require professionals, such as natural scenes annotations of pictures can be done by ordinary people. However, in life science problems, data labeling often heavily relies on professional knowledge. Therefore, labeling large-scale biological data can be more challenging. For example, in the study of protein structure, the mapping between the three-dimensional folding structure of the protein and the corresponding DNA sequence requires the expertise of structural biologists; in medical research, providing accurate disease diagnosis for pathological images requires a doctor's professional knowledge and rich diagnosis experiences. Unlike semi-supervised learning in the context of natural scenes, only a small number of experts have required expertise. Therefore, in the study of large-scale biomedical data samples, how to effectively perform data analysis becomes a critical issue.

Most semi-supervised learning strategies are based on the expansion of existing unsupervised learning or supervised learning methods by combining other additional information in the data. The first type of problem in semi-supervised learning is the classification of labeled and unlabeled data (or partially labeled data), which is an extension of the supervised classification problem. In such problems, it is generally assumed that there are more unlabeled data than labeled data. The goal of semi-supervised classification is to train a classifier on partially labeled data to achieve better performance than a classifier that is only trained on labeled data. The other type is called clustering problem with partial category constraints, which is an extension of traditional unsupervised clustering. In the existing labeled data, the samples containing the same label are known as samples in the same category; and the samples with different labels are definitely samples that are not in the same category of data. Under the constraint of known labels, the goal of constrained clustering is to obtain better clusters than clusters in only unlabeled data. Besides methods mentioned above, there are other semi-supervised learning methods and learning tasks. For example, performing regression analysis on both labeled and unlabeled data when samples are labeled as continuous values ; performing compressed feature representation on some labeled data and extract general low-dimensional effective representations.

1.2.3 Sample Missings in Biological Data

Another serious missing situation is the lack of overall sample data information. In many practical applications, a large amount of data is generated, but the data of different categories is usually unevenly distributed [21]. If the number of samples from one class is much higher than other classes, the dataset is considered as highly biased. In an imbalanced dataset, the class with more instances is called the major class, and the class with a relatively small number of instances is called the minor class. This imbalanced dataset is widely existed in life science study for a variety of reasons. The first one is that when collecting data, the minor category data is relatively rare thus it is impossible to collect as many samples as the major category. For example, when designing a spam detection algorithm for email system, only a small amount of spam data can be collected compared to non-spam data. When studying the click prediction of advertisements, most of them are unclicked, resulting to serious deviations in the data. In the field of biomedical study, this problem is particularly serious. For example, when studying rare genetic diseases, healthy samples can be collected from any groups of people, but for diseased individuals it is difficult to collect enough data. The second type of sample missing is that many samples are collected under complex and uncontrollable conditions. For example, many biological experiments need to be based on a specific environment, temperature, instrument, medicine, etc. And it is difficult to repeat the same experimental conditions. Thus the number of samples will also be highly imbalanced. The third type of sample missing is in the cost-imbalance problem. This kind of situation usually exists in the classification problem. The sample types involved in the classification may be similar in sample numbers, but the cost of misclassifying the sample is obviously different. The cost of misclassifying a certain type of sample into others is much greater than the misclassification of others. Therefore, we hope to avoid this wrong classification as much as possible. This kind of problem is most apparent in clinics. For example, in disease diagnosis, the cost of misdiagnosing a sick individual as health is much greater than the opposite scenario; in radar detection, it is always hoped that the information of the enemy aircraft will not be missed. These situations require as high as possible on the accuracy of a certain type of sample.

In the data analysis of these imbalanced data, the performance of traditional machine learning methods will be significantly degraded. The main reason is that most standard machine learning methods assume that the method is run on balanced data, with the same distribution or misclassification cost for each category. Therefore, when the data exhibits a complex and imbalanced nature, the information provided by the data itself is inconsistent with the objectives of the method. Machine learning on imbalanced datasets is called imbalanced learning, which can be mainly divided into sample sampling learning and cost-sensitive learning.

The first method directly focuses on the imbalanced number of samples in different categories, including downsampling major samples to match minor samples to form a balanced dataset; or modeling the distribution of minor sample data through statistical methods, and randomly collects minor samples from the distribution. The data will eventually form a balanced distribution. The balanced data after sampling processing can be directly combined with any traditional machine learning methods, and has good generalizability. The second method is to modify the loss function of traditional machine learning, e.g. defining a loss matrix to describe the cost of misclassification of any specific data samples and improving the mistake error of the specific sample to force the model to increase the importance of minor samples. The theoretical basis and algorithms of loss-sensitive methods can be naturally applied to imbalanced learning problems.

1.3 Main Research Content

In the context of life science research, this book explores the above three missing data types (Fig. 1.5). For each missing scenario, this book focuses on the application of life science problems and proposes different machine learning and statistical solutions. Finally, on a number of actual biological data and biological tasks, this book further verifies the effectiveness of the method and identifies new biological discoveries.

1.3.1 Statistical Reconstruction of Missing Features

Aiming at the problem of missing data feature information, this book proposes a deep recurrent self-supervised learning method. Combining functions of noise reduction and feature compression, a recurrent nerual network structure of feature imputation is proposed to effectively improve the reconstruction accuracy of missing features. At the same time, the proposed method takes into account the experimental batch errors of data measurement and achieves an end-to-end learning framework of multiple practical biological problems. Taken together, we provide an effective tool for the research of single-cell genomics, including experimental error correction, data com-

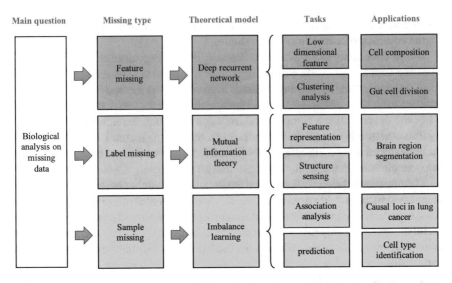

| Main question | Missing type | Theoretical model | Tasks | Applications |

Fig. 1.5 The main research contents, model theories, functions and biological applications of this book

pression, data denoising and feature missing imputations. The proposed method is applied to existing multiple biological datasets. Quantitative results evaluation shows that the proposed method can effectively impute the missing sequencing defects and improve the ability to identify unknown cell populations in biological tissues and organs. We further demonstrates the advantages of the proposed method on large-scale data and it can quickly process the genetic data of million scale in a short time and with low memory consumption.

1.3.2   Statistical Reconstruction of Missing Labels

Focusing on the problem of missing data label information, this book proposes a robust information theoretic learning method to embed the data to features and measure the degree of mutual verification between the features and the original or predicted labels of the data. With maximal mutual information from the information theory, the problems of feature learning and label prediction on partially labeled data can be effectively solved. At the same time, by combining deep learning and structured compressive sensing theory, the robust information theoretic learning method can be extended to achieve the hierarchical transformation of features and the effective use of structured prior knowledge. On the data of different degrees of incorrect and missing labels, it is verified that the proposed method can effectively improve the ability of feature learning and label prediction. Based on this, the method is further applied to the study of brain functional magnetic resonance imaging (f-MRI), and shows an high-precision segmentation on only partially tagged images.

1.3.3 Statistical Reconstruction of Missing Samples

For the missing of overall sample information, this book proposes a structure-aware rebalancing learning method. For two specific problems under this condition, including statistical inference of associations and classification on imbalanced data, we build feature structure-aware rebalancing learning and sample structure-aware rebalancing learning methods, respectively. The first method aims at the insufficient number of minor samples in the association analysis. In the framework of ensemble learning, the method samples balanced data subsets, analyzes the key features, and summarizes the conclusion of each subset and achieves an effective perception of the overall structure of data features. Then learning weights for key samples are increased to encourage the association analysis to bias towards minor samples; the second method combines deep feature transformation with sample clustering rebalancing technique, to provide an overall distribution structure of data in the context of classification learning. It simultaneously provides high quality feature transformation and balanced learning objective function. In practical biological applications, these two methods were verified to effectively identify the causal gene loci of gastric cancer and identify rare cell types in biological tissues.

The outline of this book is as follows. In Chap. 2, we introduce the statistical recovery of missing data features; in Chap. 3, we introduce the statistical recovery of missing labels; in Chap. 4, we introduces the statistical recovery of missing data sample information; finally in Chap. 5, we summarize the full text and outlook future directions.

References

1. David JB, Eric JH (2007) Computed tomography—an increasing source of radiation exposure. N Engl J Med 357(22):2277–2284
2. Sanger F, Nicklen S, Coulson AR (1977) Dna sequencing with chain-terminating inhibitors. In: Proceedings of the national academy of sciences, 74(12):5463–5467
3. Shendure J, Ji H (2008) Next-generation dna sequencing. Nat Biotechnol 26(10):1135
4. Logothetis NK, Pauls J, Augath M, Trinath T, Oeltermann A (2001) Neurophysiological investigation of the basis of the fmri signal. Nature 412(6843):150
5. Biswal B, Yetkin FZ, Haughton VM, Hyde JS (1995) Functional connectivity in the motor cortex of resting human brain using echo-planar mri. Magn Reson Med 34(4):537–541
6. Zietkiewicz E, Rafalski A, Labuda D (1994) Genome fingerprinting by simple sequence repeat (ssr)-anchored polymerase chain reaction amplification. Genomics 20(2):176–183
7. Orita M, Suzuki Y, Sekiya T, Hayashi K (1989) Rapid and sensitive detection of point mutations and dna polymorphisms using the polymerase chain reaction. Genomics 5(4):874–879
8. Hell SW, Wichmann J (1994) Breaking the diffraction resolution limit by stimulated emission: stimulated-emission-depletion fluorescence microscopy. Opt Lett 19(11):780–782

9. Huang B, Bates M, Zhuang X (2009) Super-resolution fluorescence microscopy. Annu Rev Biochem 78:993–1016
10. Adrian M, Dubochet J, Lepault J, McDowall AW (1984) Cryo-electron microscopy of viruses. Nature 308(5954):32
11. Dubochet J, Adrian M, Chang JJ, Homo JC, Lepault J, McDowall AW, Schultz P (1988) Cryo-electron microscopy of vitrified specimens. Q Rev Biophys 21(2):129–228
12. Xu M, Wang LV (2006) Photoacoustic imaging in biomedicine. Rev Sci Instrum 77(4):041101
13. Wang LV, Hu S (2012) Photoacoustic tomography: in vivo imaging from organelles to organs. Science 335(6075):1458–1462
14. Shah S, Takei Y, Zhou W, Lubeck E, Yun J, Eng CHL, Koulena N, Cronin C, Karp C, Liaw EJ et al (2018) Dynamics and spatial genomics of the nascent transcriptome by intron seqfish. Cell 174(2):363–376
15. Gut G, Herrmann MD, Pelkmans L (2018) Multiplexed protein maps link subcellular organization to cellular states. Science 361(6401):eaar7042
16. Gawad C, Koh W, Quake SR (2016) Single-cell genome sequencing: current state of the science. Nat Rev Genet 17(3):175
17. Ehud S, Tamir B, Sten L (2013) Single-cell sequencing-based technologies will revolutionize whole-organism science. Nat Rev Genet 14(9):618
18. Yuan GC, Cai L, Elowitz M, Enver T, Fan G, Guo G, Irizarry R, Kharchenko P, Kim J, Orkin S et al (2017) Challenges and emerging directions in single-cell analysis. Genome Biol 18(1):84
19. Klein AM, Macosko E (2017) Indrops and drop-seq technologies for single-cell sequencing. Lab Chip 17(15):2540–2541
20. Grün D, van Oudenaarden A (2015) Design and analysis of single-cell sequencing experiments. Cell 163(4):799–810
21. He H, Garcia EA (2008) Learning from imbalanced data. IEEE Trans Knowl Data Eng (9):1263–1284

Chapter 2
Fast Computational Recovery of Missing Features for Large-scale Biological Data

The lack of feature information is common in biological data and can seriously degrade the performance of existing data analysis methods. This chapter focuses on missing gene features in single-cell transcriptomics data. In the rapidly development of single-cell sequencing, the latest technological advances have made it possible to measure the intrinsic activity of single cells on a large scale, and enable to analyze the composition of cells within tissues with high precision. Based on this technology, many important biological structure identification methods have been proposed for the analysis of gene data. However, the missing genetic features have seriously hindered the full exploration of the internal information of biological data. For most of existing datasets, only about 20% of the genetic profiles can be effectively measured. Facing this problem, this chapter proposes deep recurrent autoencoder learning to achieve accurate and rapid imputation of missing gene expressions from millions of cell expression data.

2.1 Research Background

Single-cell RNA-seq (scRNA-seq) provides high-resolution dissection of complex biological systems, including identification of rare cell subpopulations in heterogeneous tissues, elucidation of cell-developmental programs, and characterization of cellular responses to perturbations [1–3]. Recent scRNA-seq experimental platforms [4–6] have enabled interrogation of millions of cells, offering an unprecedented resolution at which to dissect cell-type compositions.

These advances have led to two acute challenges. First, single-cell transcriptomics is susceptible to amplification noise and dropout events [7, 8], which become more pronounced as tradeoffs are made to sequence larger numbers of cells. Second, computational memory and/or speed restrictions may render analytical packages poorly scalable for large datasets [7–12]. To extract informative representations from

© Tsinghua University Press 2021
F. Bao, *Computational Reconstruction of Missing Data in Biological Research*,
Springer Theses, https://doi.org/10.1007/978-981-16-3064-4_2

extremely noisy, massive, high-dimensional single-cell RNA profiles, we developed scScope, a deep-learning-based software package. scScope uses a recurrent network layer to iteratively perform imputations on zero-valued entries of input scRNA-seq data. scScope's architecture allows imputed output to be iteratively improved through a selected number of recurrent steps.

2.2 Related Works

In the existing biological data imputation method, according to its mathematical background, it can be classified into two types: imputation under specific statistical assumptions and imputation without statistical assumptions. In the first type of methods, it is generally assumed that the way that data were generated is firstly modeled based on each stage of the biological experiment process. Then a unified inference is performed in the model framework to estimate parameters. For the method in this class specifically on single-cell sequencing data, representative approaches include ZIFA [7], and different improved versions of ZIFA, such as scVI [13], DCA [14]. ZIFA assumes that the generated data follow the zero-inflatted non-negative binomial distribution (ZINB). However, due to the complexity of the probability assumption, solution to this distribution is inefficient, which can only be solved using the traditional Bayesian probabilistic graph model , such as the Markov chain Monte Carlo method (MCMC). Especially when working with large-scale data (> 10 thousand), solving data parameters becomes a difficult problem. The second type, imputation without hypothesis, does not consider the generation factors of the data. Instead, it takes the optimal reconstruction of the known data as the learning target. The representative methods include autoencoder. But when faced with serious data missings, accurate reconstruction also becomes a difficult problem.

In this chapter, we introduce deep recurrent autoencoder learning model and other seven existing missing imputation methods: PCA, MAGIC, ZINB-WaVE, SIMILR, and three "deep" learning methods, autoencoders (AE), scVI And DCA.

2.3 Deep Recurrent Autoencoder (scScope)

The architecture of the scScope network has four modules (Fig. 2.1). The parameters in these layers are learned from data in an end-to-end manner through optimization. We note that scScope is flexible in terms of normalizing and scaling of input data, as long as the input values are non-negative.

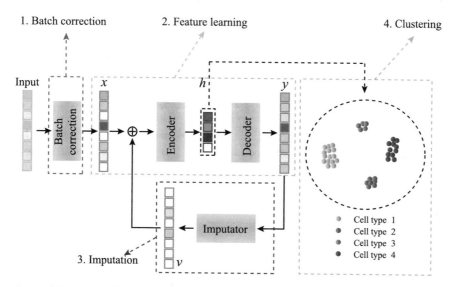

Fig. 2.1 Structure and function modules of deep recurrent autoencoder

2.3.1 Batch Effect Correction

scScope offers the option to correct for batch effects, inspired by a previously developed approach. Here, we denote: the input single-cell profile by $\tilde{x}_c \in \mathbb{R}^N$; the number of batches by K; the binary experimental batch effects indicator vector $u_c \in \mathbb{R}^K$ (nonzero entry indicates the batch of $\tilde{x}_c$); and the learnable batch correction matrix as $B \in \mathbb{R}^{N \times K}$. The batch correction layer is given by:

$$x_c = f_B = r(\tilde{x}_c - Bu_c). \tag{2.1}$$

Throughout, $r(\cdot)$ denotes the standard rectified linear unit (ReLU):

$$r(v) = v^+ = max(0, v), \tag{2.2}$$

where the ReLU enforces $x_c \geq 0$, which is expected for actual gene count data, and is widely used in deep learning. By default, $u_c = 0$ assuming a single batch.

2.3.2 Feature Learning

The feature learning module is composed of encoder and decoder two functions.

Encoder

For each cell c, the encoder layer $f_E(\cdot)$ compresses the high-dimensional batch-corrected single-cell expression profile $x_c \in \mathbb{R}^N$ into a low-dimensional, latent representation $h_c \in \mathbb{R}^M$, $M < N$. The encoder layer is given by:

$$h_c = f_E(x_c) = r(W_E x_c + b_E), \tag{2.3}$$

with learnable parameters $W_E \in R^{\in}\mathbb{R}^{N \times M}$ and $b_E \in \mathbb{R}^M$. We note that ReLU is the most widely used non-linear function in deep learning due to its ease in back-propagating gradient information across layers.

Decoder

A decoder layer is established to decompress the latent representation h_c to an output $y_c \in \mathbb{R}^N$ that is of the same dimension as the input single-cell profile,

$$y_c = f_D(h_c) = r(W_D h_c + b_D), \tag{2.4}$$

where the decoder layer $f_D(\cdot)$ is a composition of a linear transform with learnable parameters $W_D \in \mathbb{R}^{N \times M}$ and $b_D \in R^N$ followed by the nonlinear activation $r(\cdot)$. The nonlinear ReLU activation $r(\cdot)$ sets all negative values to zero, which makes experimental sense for gene transcript abundance measured by nonnegative values. When minimizing the differences between y_c and x_c ($\forall c = 1 \ldots n$) in the training set (of size n), the above encoder-decoder network acts as a typical auto-encoder (AE) neural network. AEs exhibit the potential for generating a "clean" signal y_c by removing additive noise from x_c via L_p norm minimization (typically, $p = 2$). However, the AE paradigm lacks a mechanism for dealing with dropout entries (missing data), which can be an even more severe problem than noise for scRNA-seq.

2.3.3 Imputation

We next focus on estimating missing data measurements. Zero-valued measurements can be due either to biological reasons (un-transcribed genes), or to technical reasons (dropouts). Our imputation step is designed to deal with the later problem, namely to estimate dropout gene expression values. Of course, a challenge shared by all single-cell-based imputation is to distinguish between zero-expressed vs. dropout measurements.

To this end, we developed a self-correcting layer to impute missing entries caused by dropout during sequencing. Our imputation approach implicitly makes use of the fact that subsets of genes are often co-regulated (e.g. by common transcription factors or pathway activation) and that their patterns of co-expression can be learned by observation of sufficiently many cells. The decoder in our AE framework is enabled to find such patterns from the latent space representation. The studies of transcriptional regulatory mechanisms have led to the development of many approaches for network-based pathway recovery.

For our purposes of ecoder layer output y_c to a p-dimensional latent vector

$$u_c = r(W_U y_c + b_U) \in \mathbb{R}^p, \tag{2.5}$$

with learnable parameters $W_U \in \mathbb{R}^{p \times N}$ and $b_U \in \mathbb{R}^p$, $p < N$ (we set $p = 64$). Then, we performed imputation by:

$$v_c = f_I = P_{Z_c}[r(W_V u_c + b_V)] \in \mathbb{R}^N, \tag{2.6}$$

with learnable parameters $W_V \in \mathbb{R}^{p \times N}$ and $b_V \in \mathbb{R}^N$. Here, Z_c (resp. $\overline{Z}_c$) is the set of zero- (resp. non-zero-) valued genes in profile x_c of the c-th cell, and the entry-sampling operator P_{Z_c} sets all entries not in Z_c to zero (i.e., for vector r, with entries r_j, $P_{Z_c}(r_j) = r_j$ if $j \in Z_c$ and $P_{Z_c}(r_j) = 0$ if $j \notin Z_c$. We used Z_c as we are only interested in imputation for zero-valued genes in the profile x_c.

After obtaining the imputed vector v_c, we calculated the corrected single-cell expression profile: $\hat{x}_c = x_c + v_c$. This corrected single-cell profile $\hat{x}_c$ was then re-sent through the encoder-decoder framework to re-learn an updated latent representation.

2.3.4 Learning the Objective Function

The imputation layer imposes a recurrent structure on the scScope network architecture. For clarity of exposition, the recurrent scScope can be unfolded into multiple time steps (Fig. 2.2 shows three steps). Then, the whole recurrent scScope framework can be described as:

$$h_c^t = f_E(\hat{x}_c^t) = f_E(x_c + v_c^{t-1}), \ y_c^t = f_D(h_c^t), \ v_c^t = f_I(y_c^t), \ v_c^0 = 0, \tag{2.7}$$

for iterations $t = 1 \ldots T$. At the first step, the correcting layer's output v_c^0 is set as zero. For $T = 1$, scScope is a standard auto-encoder (AE).

The learning objective for scScope is defined by the pursuit of unsupervised, self-reconstruction (as typically used in AE training):

$$f_B, f_E, f_D, f_I = argmin \ L = \sum_{c=1}^{n} \sum_{t=1}^{T} \| P_{\overline{Z}_c}[y_c^t - x_c] \|^2 \tag{2.8}$$

The entry-sampling operator $P_{\overline{Z}_c}$ forces loss computation only on non-zero entries of x_c. The parameters in the batch correction layer (f_B), encoder layer (f_E), decoder layer (f_D) and imputation layer (f_I) are all learned by minimizing the above loss function.

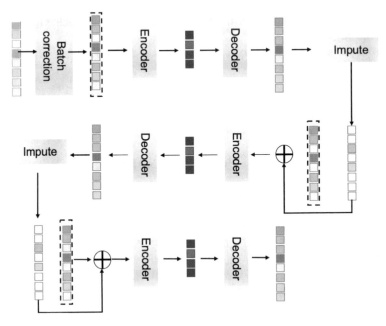

Fig. 2.2 Unfold the recurrent neural network structure

2.3.5 Multiple GPU Training

scScope offers the option to train its deep model by using multiple GPUs, which can dramatically reduce runtime (Fig. 2.3). In this mode, scScope replicates its network structure on multiple GPUs and aggregates all network parameters on the CPU. These network parameters include the connections and biases of all encoder, decoder and imputation layers of scScope. In a round of batch training, one GPU grabs the current network parameters from the CPU to use for its own network replicate of scScope. Then, for gradient calculation, the GPU processes a randomly chosen batch of (= 64 or 512) single-cell expression profiles from a total of single-cell profiles. We apply a conventional gradient calculation framework for neural networks, which iteratively performs feed-forward and back propagation steps. In the feed-forward (FF) step, a GPU passes its batch of single-cell samples through its locally stored scScope network and accumulates the losses for this batch. In the back propagation (BP) step, batch-dependent gradient information for network parameters on different layers is calculated by sequentially propagating accumulated loss from the end to the first network layer. This BP operation is performed by using gradient calculation functions wrapped in deep-learning packages (in our case TensorFlow). We apply this process independently across all k GPUs in a parallelized manner to obtain gradient information from a total samples. The gradient information of those GPUs is averaged by the CPU, i.e. $G^{(j)} = (G_1^{(j)} + \ldots + G_k^{(j)})/k$, where $G_k^{(j)}$ is the gradient calculated from the GPU in the jth round of optimization iteration. Finally, we apply

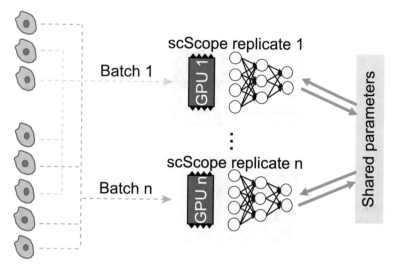

Fig. 2.3 Training of scScope is implemented on multiple GPUs to enable fast learning and low memory cost

adaptive moment estimation (ADAM) with default TensorFlow parameters to update the network parameters stored on the CPU. Iterations were terminated when either the objective function showed little change (i.e. $< 0.1\%$) or the number of iterations reached a maximal epoch (e.g. 100).

2.3.6 Cell Subpopulation Discovery

Representations outputted by scScope at each step can be concatenated as a long feature vector, which is easily integrated with any clustering method. We used the graph-based method PhenoGraph [15], as it performs automated robust discovery of subpopulations, as well as determines subpopulation numbers automatically.

Graph clustering for moderate-scale data

We directly applied the PhenoGraph software to datasets of moderate scale. All clustering results were obtained using a Python-implemented PhenoGraph package (version 1.5.2). We followed the suggested setting and considered 30 nearest neighbors when constructing graphs.

Scalable graph clustering for large-scale data

scScope enables the feature learning on millions of cells. However, PhenoGraph is unable to handle millions of cells due to the extreme computational costs and memory requirements in graph construction. To leverage the power of graph clustering on analyzing these large-scale data, we designed a density down-sampling

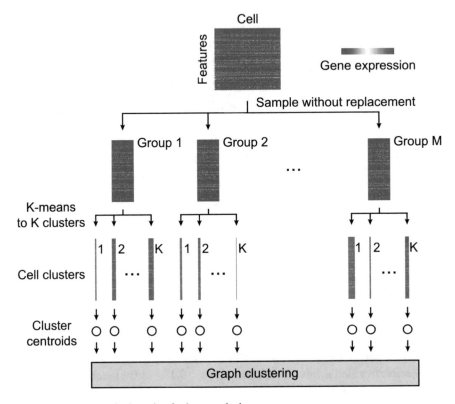

Fig. 2.4 Scalable graph clustering for large-scale data

clustering strategy by combining k-means and PhenoGraph. In detail, we divided cells into M groups with equal size and performed k-means clustering on each group independently (Fig. 2.4). The whole dataset was split to $M \times K$ clusters and we only input the cluster centroids into PhenoGraph for graph clustering. Finally, each cell was assigned to graph clusters according to the cluster labels of its nearest centroids. In our implementation on the dataset of 1.3 million mouse brain cells, we took $M = 500$ and $K = 400$, which made it possible to process millions of data in only tens of minutes without loss of accuracy.

Scalable memory allocation for analyzing large numbers of genes in large datasets

For large datasets and gene numbers, scScope implements a scalable memory allocation strategy that allows the dataset to be broken into a smaller number of batches that can be loaded directly into memory. We note that when this option is used, minibatches are only selected from within each batch during training. This option was only used for the case of $\geq 10K$ genes; here the $400K$ mouse cell atlas dataset was broken into four batches of size $100K$.

Table 2.1 Summary of comparing methods

	Method type	Imputation	Batch correction	GPU parallel training
PCA	SVD	–	–	–
AE	Neural network	–	–	Yes
MAGIC	Linear algebra	Yes	–	–
ZINB-WaVE	Matrix factorization	Yes	Yes	–
SIMLR	Kernel learning	–	–	–
DCA	Deep neural network	Yes	–	–
scVI	Bayesian inference	Yes	Yes	–
scScope	Recurrent autoencoder	Yes	Yes	Yes

2.4 Experiments

2.4.1 Comparing Methods

We note that a hyper-parameter across all methods is the number of latent feature dimensions M; we observed that all methods were reasonably robust to changes in M; and to avoid an intractable number of possible $M = 50$ for all comparisons. Unless otherwise noted, the software packages were used as follows (Table 2.1).

1. **MAGIC**: The MAGIC algorithm (Markov Affinity-based Graph Imputation of Cells, MAGIC) was performed using the python-based package magic. We input the raw data and employed the function provided by the software for all simulated and real data to learn 50-dimension latent features.
2. **ZINB-WaVE**: ZINB-WaVE (Zero-Inflated Negative Binomial-based Wanted Variation Extraction) employed the R package zinbwave, and all default parameters were used to learn the 50-dimension feature vector. For running without batch correction, we set batch information for each cell as the same, and for batch correction mode the batch indices were input as reference.
3. **SIMLR**: SIMLR (Single-cell interpretation via multi-kernel learning, SIMLR), We used the Python implementation of SIMLR with the authors' default parameter settings. SIMLR needs to take the desired cluster number as input. For our simulated dataset, we input the known cluster numbers. For the retina dataset, where the "true" cluster number is unknown, we set it to 39, which is the cluster number reported in the original study
4. **DCA**: DCA is an unpublished software based on TensorFlow framework. We installed the Python package of DCA (download date: Sep. 4, 2018) and ran DCA by setting the latent dimension to 50. We set the training epoch to 100 and kept all other default parameters.
5. **scVI**: scVI (Single-cell Variational Inference), We used the Torch-based Python package of scVI (download date: June 5, 2018). In the original demonstration of

the software, they assigned different parameters for different test datasets. Here, we set the training step to 0.001, epoch to 100 and the latent dimension to 50. The software shuffled the cell order randomly in training and did not offer the option to output the latent representation with the same input cell orders. In order to keep track of the input cells for later analysis, we appended cell IDs to the cell labels.

6. **scScope**: scScope was implemented in Python 3.6 with TensorFlow-GPU 1.4.1, Numpy 1.13.0, Scikit-learn 0.18.1 packages and was tested on a server with 4 GPUs (Nvida Titan X) and 64GB memory. For all experiments presented in this paper, we extracted 50-dimensional representations with 2 recurrent learning steps $(T = 2)$. We use the default batch size of 64 and epoch of 100 in training. Unless noted specifically, scScope was run on single GPU mode in our comparisons.

Down-sampling training strategies on large-scale dataset: It is not possible to run directly the non-deep-learning based approaches on dataset with millions of cells. For these packages, we randomly down-sampled datasets containing more than 1M cells to a subset of 20K cells. On these down-sampled datasets, single-cell feature vectors were learned by the respective method and clustered by PhenoGraph. A support vector machine (SVM) was trained on this subset in the latent feature space and then used to assign labels for the rest of cells in the unsampled dataset. The deep-learning approaches we tested can learn features on millions of cell profiles, but the software does not provide a function for automatic clustering on such large-scale datasets. Therefore, for comparisons we passed the output of their deep learning algorithms for single-cell feature learning to our scalable graph clustering approach for large-scale clustering.

All compared methods were run on the same server with Xeon E5 CPU, 64 GB memory, Nvidia Titan X GPU and Ubuntu 14.04 operation system. Further, all comparisons were performed using log transformed input.

2.4.2 Evaluation of Clustering and Batch Correction

For the proposed method, its performance is evaluated from multiple perspectives, including the ability to fill missing data, the ability to make use of features, and finally the ability to analyze the structure of data.

Performance of clustering

We used the adjusted Rand index (ARI) to compare label sets of two clustering methods. For two clustering results U and V with r and s clusters on a data set of n cells, n_{ij} denotes the number of cells shared between cluster i in U and cluster j in V. And ARI is defined as

$$ARI = \frac{\sum_{ij}\binom{n_{ij}}{2} - \left[\sum_i \binom{n_{i*}}{2} \sum_j \binom{n_{*j}}{2}\right] / \binom{n}{2}}{\frac{1}{2}\left[\sum_i \binom{n_{i*}}{2} + \sum_j \binom{n_{*j}}{2}\right] - \left[\sum_i \binom{n_{i*}}{2} \sum_j \binom{n_{*j}}{2}\right] / \binom{n}{2}},$$

$$(2.9)$$

where $n_{i*} = \sum_j n_{ij}, n_{*j} = \sum_i n_{ij}$, and n is the number of cells in the data set.

Evaluation of Imputation

The imputation accuracy was defined as the normalized distance between the imputed log count entries and log count ground truth entries. We constructed lists l and $\hat{l}$, whose elements correspond to either ground truth or imputed values (respectively) for all dropouts entries across all cells. We defined the normalized error as:

$$error = \frac{\| l - \hat{l} \|_1}{\| l \|_0},$$

$$(2.10)$$

where $\| \cdot \|_p$ means the p norm of a vector.

For real biological data, the ground truth values for missing genes are unknown. To evaluate scScope's imputation accuracy on real biological data, we followed the same down-sampling strategy as used for scVI. Namely, we randomly split the entire collection of n cells into n_{train} training cells and n_{val} validation cells. We used the different imputation methods to build gene models from the n_{train} cells. On each of the n_{val} cells, we randomly set $p\%$ of its non-zero genes as "simulated" missing genes and set their corresponding count values to zero. The real measured values of these simulated missing genes were then used to generate the ground truth list l, and the list $\hat{l}$ was based on inferred values for the simulated missing genes from the $n_{v}al$ cells. The reconstruction error was calculated as for simulated data above.

2.4.3 Generation of Simulation Data

In order to quantitatively evaluate the phenotype of the method and compare the performance of various methods, firstly, different simulation methods are used to generate simulation data. In the simulation, all gene expression data can be directly observed. This is the main difference from the real measurement data. Therefore, in the reconstruction analysis, some data is randomly assumed to be missing, and the real missing data is known. You can use this to judge the results.

Simulated data by Splatter

The simulation package Splatter [16] is designed to generate realistic scRNA-seq data. We used this package to generate data with 2000 cells, 3 subpopulation groups, and dropout rates from 1 to 5.

Simulated data by SIMLR

We used the simulation approach from SIMLR12 to generate large-scale scRNA-seq data due to limitations in scalability of Splatter. Following previous studies, we initially tested the performance of scScope for cell-subpopulation discovery using simulated data. We assumed the high-dimensional single-cell expression data $x_c \in \mathbb{R}^N$ is generated or controlled by a latent code $z_c \in \mathbb{R}^P (P < N)$, which is not observable. z_c is sampled from a Gaussian mixture model with k Gaussian components, *i.e.* $z_c \sim \sum_{j=1}^{k} \pi_j N(\mu_j, \Sigma_j)$. The mixture coefficients π_j were chosen from a uniform distribution and normalized to sum to 1, the mean vector $\mu_j \in \mathbb{R}^P$ was uniformly sampled in $[0, 1]^P$, and the covariance matrix $\Sigma_j \in \mathbb{R}^{P \times P}$ was chosen to be $\Sigma = 0.1 \times I$, for identity matrix I.

To simulate single-cell gene vectors, we generated a projection matrix $A \in \mathbb{R}^{N \times P}$ to map the low-dimensional latent code to a high-dimensional space. First, we simulated ground truth, x_c^{true}, which cannot be observed:

$$x_c^{true} = Az_c + b, \qquad (2.11)$$

where each entry in A was independently sampled from the uniform distribution $U(-0.5, 0.5)$ and bias $b = 0.5$ is added to avoid negative gene expression in the high-dimensional mapping. Second, we simulated the observed profile, x_c^{obs}, which may contain gene-specific noise and dropout artifacts due to the sequencing technique and platform. Noise was added to the true gene profile by: $x_{cg}^{noise} = x_{cg}^{true} + \sigma_{cg}$, where $\sigma_{cg} \sim N(0, \Sigma_g^{noise})$ and Σ_g^{noise} was uniformly sampled in the range of $[0, 0.2]$ independently for each gene. Dropout events were added via a double exponential model with decay parameter:

$$x_{cg}^{obs} = x_{cg}^{noise} \delta[q_{cg} > \exp{-\alpha x_{cg}^{noise^2}}], \qquad (2.12)$$

where x_{cg} denotes the gth gene of x_c, q_{cg} was randomly sampled in $[0, 1]$, and $\delta = 1$, if its argument is true and $= 0$ otherwise. This double-exponential model is motivated by the widely-used assumption that low-expressed genes have higher probability to be influenced by dropout events. We use the aforementioned generative model to create N single-cell profiles $x_i^{obs}, i = 1 \ldots N$. In the reported results, we repeat such random generation process for 10 times on $n = 10K$ and 1M simulated single-cell profiles under various conditions for performance comparisons among different approaches.

Simulation with rare cell subgroups

To generate cell subpopulations with rare cell types, *i.e.* cell clusters with very limited numbers of cells compared to the major clusters, we sample the mixture coefficients π_j from a non-uniform distribution as

$$\pi_{1,\ldots,k_m} = q, \pi_{k_m+1,\ldots,k} = \frac{1 - q \times k_m}{k - k_m} \qquad (2.13)$$

Table 2.2 Parameter settings for Splatter simulation

Dropout parameter	1	2	3	4
Dropout rate	0.23	0.33	0.47	0.77

Table 2.3 Parameter setting for SIMLR simulation

Dropout coefficient	1	0.5	0.3	0.15
Dropout rate	0.61	0.73	0.78	0.88

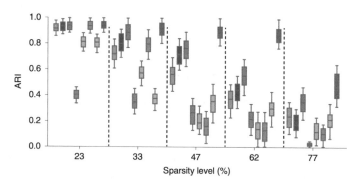

Fig. 2.5 The comparison methods find the correct rate of the sub-cell category on the data generated by the Splatter method

where $q \ll 1/k$ is the mixture fraction for each minor cluster, k_m is the number of rare cell subpopulations, and k is the total number of subpopulations. With this minor modification from the previous section, we generate imbalanced cell-type compositions with k_m rare cell types. 10 replicates for each condition were analyzed. We randomly down-sample these 1M single cell profiles to 10K. On the subsampled datasets, after learning features, we applied PhenoGraph for *de novo* cell subpopulation discovery. On the un-subsampled datasets, we used our scalable graph clustering approach. We note that subpopulation numbers are automatically determined by these clustering approaches.

2.4.4 Method Evaluations

Performance evaluation on simualted data

To calibrate the accuracy of scScope on simulated datasets, we made use of two third-party packages for generating scRNA-seq data (Tables 2.2, 2.3). First, we used Splatter to generate moderate-sized datasets of varying sparsity levels (percentage

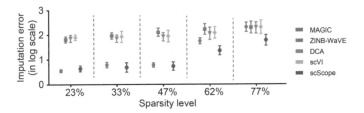

Fig. 2.6 Comparison of imputation accuracy on 2K single-cell datasets generated by Splatter

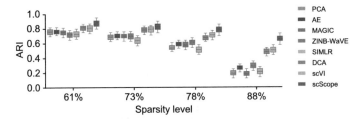

Fig. 2.7 The comparison on discovering cell clusters on the simulated data generated by the SIMLR method

of 0-valued genes), containing: 2K scRNA-seq profiles with 500 genes, and three underlying subpopulations. In terms of discovering these underlying subpopulations, scScope performs similarly to other approaches when there are only minor dropout effects, but shows a large advantage in accuracy as dropout rates increase to realistic ranges observed in biological data (Fig. 2.5).

In terms of imputation error, at low sparsity ($< 50\%$) scScope and MAGIC outperformed all other methods, though at high sparsity scScope outperformed all other approaches (Fig. 2.6).

It is pointed out here that Splatter is a very complex simulation framework, which contains very complicated statistical distribution assumptions. Therefore, when generating simulation data, due to the constraints of computing power and memory size, it is impossible to generate high-gene dimensional, large-scale cell data. Therefore, at the same time consider a simple but scalable data simulation method SIMLR, this method comes from Nature Methods.

We used the simulation framework in SIMLR to generate massive-sized and more heterogeneous datasets of varying sparsity levels containing: 1M scRNA-seq profiles with 500 genes, and 50 underlying subpopulations. To perform our evaluations, the deep approaches were able to operate over the full datasets, while the non-deep single-cell packages required down-sampled training strategies (Methods). We found that scScope performed well, particularly at high sparsity levels (Fig. 2.7).

An increasingly important task for scRNA-seq profiling approaches is to identify rare cell subpopulations within large-scale data, In order to test the ability of various methods to discover existing cell populations, simulation data containing sub-cell

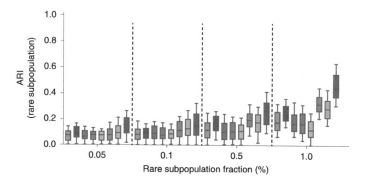

Fig. 2.8 The comparison on discovering rare cell clusters on the simulated data generated by the SIMLR method

populations of different rarities are further generated here. In this type of analysis, the experiment considers only one-tenth of the cells.

As might be expected, methods that did not require down sampling are better able to detect rare cell subpopulations. Overall, scScope performed reasonably well on this challenging task. We can even find subgroups that account for 0.05% of the overall data set (Fig. 2.8). Our calibration suggested that scScope can efficiently identify cell subpopulations from complex datasets with high dropout rates, large numbers of subpopulations and rare cell types.

Validation of true data

We next evaluated scScope on four experimental single-cell RNA datasets containing varying degrees of biological "ground truth". These datasets were used to test the ability of scScope to: remove batch effects, recover dropout genes, identify minor subpopulations, and test clustering accuracy for a large dataset with varying numbers of analyzed genes (mouse cell atlas).

The analysis of the biological data is summarized as follows:

1. **Lung tissue data**: The lung tissue dataset is part of the Mouse Cell Atlas data8. This dataset was downloaded from Gene Expression Omnibus (GEO) database (accession number: GSE108097). In this dataset, 6,940 cells were sequenced via three independent experiments by Microwell-seq, with 2,512, 1,414 and 3,014 cells in each batch. 1,000 high variable genes were selected for analysis. Then three batch correction methods (ZINB-WaVE, scVI and scScope) were used on the normalized data to learn the low-dimensional features. Cell types were identified by PhenoGraph clustering. To evaluate the clustering accuracy, identified cell labels were compared with previously reported labels (https://satijalab.org/seurat/mca.html).

2. **Cord blood mononuclear cells dataset**: In this dataset, 8,617 cord blood mononuclear cells (CBMC) were profiled by CITE-seq, a new technology which enabled the simultaneous measurement of protein levels and transcriptome levels for each cell. This dataset was downloaded from GEO database (accession

Table 2.4 Cell label genes of the retinal cell dataset

Cell type	Label genes		
Horizontal cells	*Lhx1*	*Pax6*	
Retinal ganglion cells	*Slc17A6*	*Pax6*	
Amacrine cells	*Pax6*	*Gad1*	*Slc6a9*
Cones	*Opn-millionw*		
Bipolar cells	*Vsx2*		
Muller glia	*Pax6*	*Vsx2*	*Rlbp1*
Astrocytes	*Rlbp1*	*Gfap*	
Fibroblasts	*Pax6*	*Slc6a9*	*Vsx2*
Vascular endothelium	*Pecam1*	*Kcnj8*	
Pericytes	*Kcnj8*		
Microglia	*Cx3cr1*		

number: GSE100866). Cell types in CBMC have been extensively studied and identified. Based on this prior knowledge, 13 monoclonal antibodies were chosen to identify bone fide cell types. These antibodies serve as an "orthogonal" ground truth to evaluate analysis results based on RNA-seq data. In the data pre-processing stage, we removed the spiked-in mouse cells and only kept 8,005 human cells for analysis using the cell-filtering strategy introduced in original study. The top 1,000 most variable human genes were selected for downstream analysis after the log1p transformation and normalization by library size. For antibody data, we used the centered log ratio transformed antibody-derived tags (ADTs), which is also provided by authors. To evaluate the performance of each method, we first automatically identified cell populations based on ADTs data using PhenoGraph. Then, scRNA-seq data were input to each method to learn latent representations which were used by PhenoGraph to predict cell types. The ADTs-derived cell types were taken as ground truth to evaluate the accuracy of cell types by scRNA-seq data.

3. **Retina data**: In this dataset, 44,808 cells were captured from the retinas of 14-day-old mice and sequenced by DropSeq. Data were obtained from the GEO database (accession number: GSE63473). In order to be comparable with published results, we followed previous experimental procedures to select 384-most variable genes and then to log transform their expression (log(TPM + 1)). After clustering, we annotated clusters obtained by each method using the same maker genes in the original study (Table 2.4). We identified candidate cell types based on the highest average type-specific marker expression. For each cluster, we calculated the fold-change values of all cell-type markers, and if at least one of a type-specific gene marker was expressed significantly higher (log2 fold change > 0.5) than in all other clusters, we assigned the cluster with the candidate cell type. Otherwise the cluster was assigned to the cell type "Rod cell".

Fig. 2.9 Batch error
correction performances
from different methods

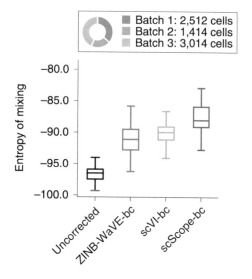

4. **Mouse cell atlas data**: The mouse cell atlas (MCA) dataset is designed to offer a comprehensive investigation of all major cell types in mouse. Data were downloaded from the GEO database (accession number GSE108097). In the dataset, 405,796 cells were sampled from 51 tissues and were sequenced by Microwell-seq. Data were firstly normalized by library size and 1,000, 2,000, 5,000, 10,000 and 20,000 top-variable genes were selected to test the scalability of each method on gene numbers. Only the deep-learning based methods (DCA, scVI and scScope) could be applied directly to this large-scale dataset. Further, to identify clusters in the MCA dataset, we applied our scalable clustering approach to the latent features. In most of the 51 tissues, one major cell type dominated the cell population. Therefore, we used the tissue identify as a proxy for ground truth to evaluate cell-type discovery.

To test the ability to remove batch effects, we made use of the lung tissue dataset (part of the mouse cell atlas), which contained 7K scRNA-seq profiles obtained from three different batches. We examined the 1,000 most variable genes and used Pheno-Graph22 to identify subpopulations. ZINB-WaVE, scVI and scScope incorporated methodology for removing batch effects, and we tested their accuracies—with or without batch effect correction—for discovering 32 previously reported cell subpopulations. To evaluate the performance of batch correction, we made use of the score developed in mutual nearest neighbors (MNNs) In brief, 100 cells were randomly sampled from the entire population, the entropy of the distributions of batches for the nearest 100 neighbors was calculated and the process was iterated 100 times and shown as box plots (Fig. 2.9). At the same time, in order to evaluate whether correcting experimental batch errors can help improve the performance of the method, the previously discovered 32 cell sets were evaluated as the true value of the analysis. When the batch correction option is enabled, all three methods improve accuracy;

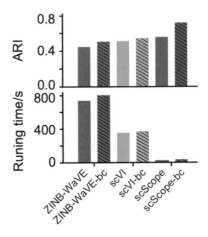

Fig. 2.10 Correcting experimental batch errors can help improve clustering performance

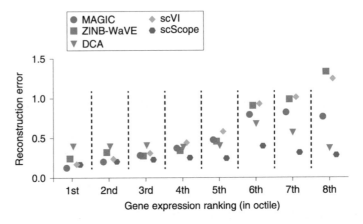

Fig. 2.11 The imputation accuracy of the comparing method on the features in different expression levels

scScope, as well as the other methods, showed improved accuracy when the option for batch correction was enabled (Fig. 2.10, left panel). Reassuringly, correcting for batch effects did not compromise the short runtime of scScope (Fig. 2.10, right panel).

What is the dependency of imputation accuracy on gene expression level? We made use of the cord blood mononuclear cell (CBMC) dataset, which contained 8K scRNA-seq profiles and 1,000 most variable genes. Following the strategy used for scVI (Methods), we sequentially simulated dropouts for genes based on octile of expression ranking (Fig. 2.11) We found for reconstructing small count values that MAGIC and scVI performed well, while for large count values DCA worked well. scScope showed small imputation errors consistently across the entire range of expression.

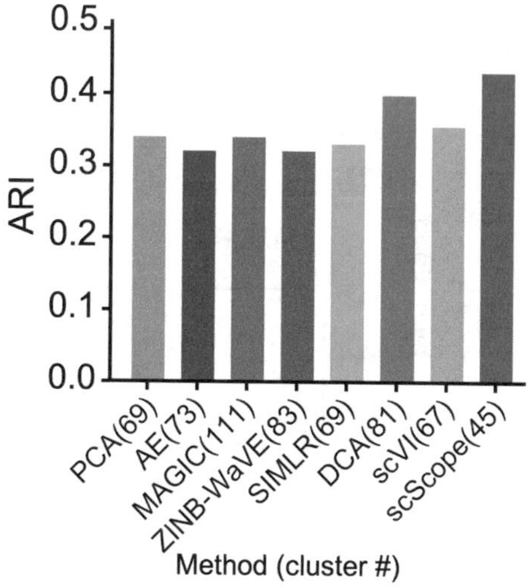

Fig. 2.12 The accuracy of different methods in identifying different cell types on retinal data

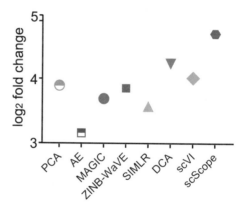

Fig. 2.13 Enrichment level of marker genes for different cell types for the retina data

Can minor subpopulations be identified in biological data? We made use of the mouse retina dataset, which contained 44K cells from dissociated mouse retinas. We applied all methods to the 384 most variable genes and used PhenoGraph to identify subpopulations The original study identified 39 cell subpopulations using fine-grained manual adjustment and expert visual inspection, which we took as a reference for our comparisons. scScope automatically identified the most similar clustering (number and assignment) to those reported in the original study (Fig. 2.12).

We annotated the clusters to cell types (Table 2.4) based on gene markers reported in the original study. After identifying the set, the expression level of each set for the corresponding biomarker further demonstrates the credible scale of each method for finding cells. If the gene marker corresponding to each cell shows a higher degree of expression enrichment, it means that the method has a better ability to find cell

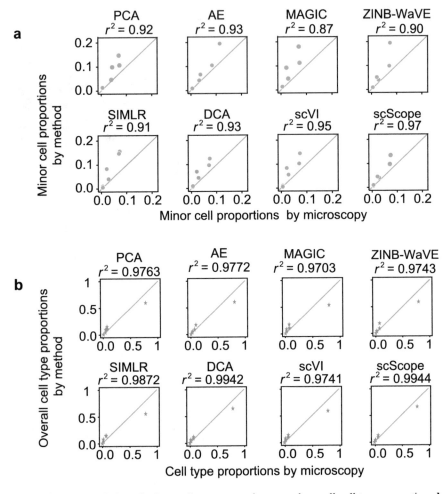

Fig. 2.14 The correlation of minor cell-type proportions **a** and overall cell-type proportions **b** identified by computation or microscopy

differences. Overall, clusters identified by scScope showed the most statistically significant enrichment of specific cell-type markers (larger fold-changes) (Fig. 2.13).

Microscopy estimates of cell-type composition and proportion either without (Fig. 2.14b) or with (Fig. 2.14) including the major cell type of rod cells. Additionally, when analyzing "pure" cell types within the Retina dataset, scScope can maintain the underlying simple structure and achieve high reconstruction accuracy on the original data.

In the previous methods, the genes in the cells were selected first, then down-sampling was performed and finally analyzed. Obviously, more genes will provide more effective information, what is the benefit of analyzing increasing numbers of genes? We made use of the mouse cell atlas, which contained 400K cells sampled

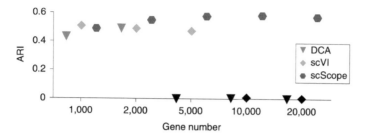

Fig. 2.15 The effect of gene number on cell type identification

Fig. 2.16 Comparison of run time

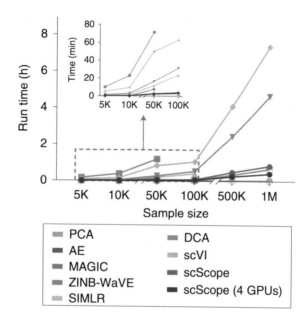

from 51 tissues. Only the deep learning algorithms were able to scale to these data sizes. To perform automatic identification of subpopulations on large datasets, we designed a scalable clustering approach (Fig. 2.4). We made use of the 51 known tissue types to assess accuracy of the clustering results. Here, we found that all three algorithms performed similarly for 1,000 and 2,000 genes (Fig. 2.15). However, the best performance overall was achieved by scScope by analyzing 10,000 genes. We note that for this analysis, we made use of an option in scScope's software for scalable memory allocation. This feature in scScope offers the opportunity to analyze increased numbers of genes.

Table 2.5 The marker gene of mouse intestinal cells

Cell type	Marker gene
Stem	*Lgr5, Ascl2, Slc12a2, Axin2, Olfm4, Gkn3*
Cell cycle	*Mki67, Cdk4, Mcm5, Mcm6, Pcna*
Enterocyte progenitor	*Alpi, Apoa1, Apoa4, Fabp1*
Enterocyte proximal	*Fabp1, Lct, Cbr1, Ephx2, Gstm3, Adh6a, Creb3l3*
Enterocyte distal	*Mep1a, Fgf15, Clec2h, Fabp6, Mep1a, Muc3*
Goblet	*Muc2, Tff3, Agr2*
Paneth	*Lyz1, Defa17, Defa22, Defa24, Ang4*
EEC	*Sox4, Neurog3, Neurod2, Sct, Cck, Gcg*
Tuft	*Dclk1, Trpm5, Gfi1b, Il25, Lrmp, Cd24a*

2.4.5 Comparison of Run Time

To analyze the performance of the methods, we tested the scalability and training speed of scScope on a mouse brain dataset, which contained 1.3M cells. We evaluated the computational costs of all methods over a wide range of subsampled data sizes (from 5K to 1M) (Fig. 2.16). As expected, the generic machine-learning tool PCA was the fastest. scScope was able to complete its analysis of the full dataset in under an hour, which was comparable to the AE, and this runtime was significantly dropped by using the option for multiple GPU training. The non-deep single-cell software packages were unable to scale beyond 100K cells, and the deep approaches, while able to scale to 1M cells, required at least seven times more computing time than scScope. Thus, scScope offers a scalable and highly efficient approach for analyzing large scRNA-seq datasets.

2.5 Biological Application: Construction of a Large-scale Biological Organ Cell Map

Finally, we applied scScope to investigate novel biology in datasets. We focused on the ability to reveal changes in cell-type composition under perturbed conditions (Chap. 2.5.2), and the ability to scale to large datasets and reveal new subpopulations (Chap. 2.5.3).

2.5.1 *Introduction of Analysis Data*

Data on different infection conditions of mouse intestines

In this dataset, intestinal epithelial cells were captured from 10 mice and sequenced using droplet-based scRNA-seq. Data were downloaded from the GEO database (accession number GSE92332). Among all cells, 1,770 cells from 2 mice were infected by *Salmonella* for 2 days; 2,121 cells (2 mice) and 2,711 cells (2 mice) were infected by *H. polygyrus* for 3 and 10 days, respectively. An additional 3,240 cells were sequenced from 4 healthy mice as a control group. We again followed the same procedure that log-transformed the expression data and selected top 1,000 most variable genes as input to scScope. For cell subpopulation annotation, we first assigned clusters to one of 7 major cell types (stem, cell-cycle related, distal enterocyte, proximal enterocyte, goblet, enteroendocrine, and tuft) according to the maximum averaged expression of cell-type makers (Table 2.5). Second, cell-cycle related clusters were subdivided into increasing stages of maturation (transit amplifying early stage, transit amplifying late stage, and enterocyte progenitor) based on the ratio of cell-cycle & stem cell markers to enterocyte expression. Third, the distal and proximal enterocyte clusters were further classified (immature vs. mature) based on increasing expression levels of the enterocyte gene markers. After annotating clusters, we calculated the cell proportion for each mouse and then averaged the proportions among mice of the same infection condition. For significant tests of proportion changes after infection, we compared proportions of mice in control group and in infection group using a two-sided t-test and rank-sum test. P-values were obtained under the null hypothesis that no changes happened in proportions after infection. Overexpressed genes for each cluster were also identified by the same differential expression analysis.

Brain data

Data were obtained from 10x Genomics (http://10xgenomics.com). 1,308,421 cells from embryonic mice brains were sequenced by Cell Ranger 1.3.0 protocol. We transformed unique molecular identifier (UMI) count data into log(TPM+1) and calculated the dispersion measure (variance/mean) for each gene. According to the rank of the dispersion measure, we selected the top 1,000 most variable genes for analysis. Due to the massive scale of the dataset, we set the batch size of scScope to 512 and trained the model for 10 epochs. Cells were further clustered into 36 groups by our density down-sampling clustering. We annotated clusters to three major types (excitatory neurons, inhibitory neurons and non-neuronal cells) based on maximal-expressed maker genes. To identify cluster-specific overexpressed genes, we then conducted differential expression analysis for each gene. We normalized UMI-count to the range of [0 1] for each gene, enabling comparisons across genes. Then gene-expression fold-change and rank-sum P-values were calculated between cells within vs. outside each cluster. Significantly overexpressed genes were identified using the criteria of $\log_2$ fold change > 0.5 and rank-sum P-value < 0.05. Data set was download from https://support.10xgenomics.com/single-cell-gene-expression/

datasets/1.3.0/1M_neurons on December 10, 2017. The data analysis by 10xge-monics was obtained from http://storage.pardot.com/172142/31729/LIT000015_Chromium_Million_Brain_Cells_Application_Note_Digital_RevA.pdf.

2.5.2 Biological Discovery of Mouse Gut Data

The intestinal dataset contained ∼10K cells obtained from mouse intestines with different infection conditions. We applied scScope to the 1,000 most variable genes (Fig. 2.17a). In the original study, enterocytes were identified as a single cluster.

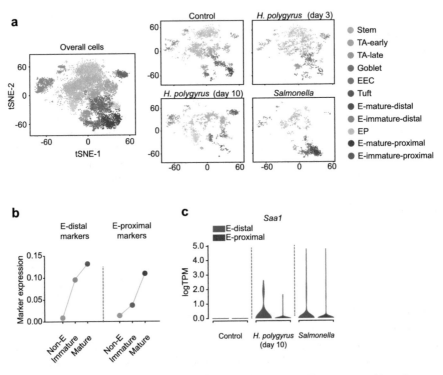

Fig. 2.17 Analysis of intestinal scRNA-seq dataset. **a** Changes in cell-type composition of mouse intestinal epithelial under different infection conditions, visualized via tSNE plots (n=9,842 cells in total). TA: transit amplifying, EEC: enteroendocrine, EP: enterocyte progenitor, E: enterocyte. scScope identified four subtypes of enterocyte cells. **b** Identification of mature versus immature and distal versus proximal enterocyte subpopulations. Shown are expression levels of E-distal and E-proximal gene markers (average UMI count) on the four enterocyte subtypes predicted by scScope and, for comparison, all other clusters (Non-E). **c** Discovery of differential expression of the gene *Saa1* in distal and proximal enterocytes after *Salmonella* and *H. polygyrus* infections. Violin plot shows the distributions of logTPM for E-distal/E-proximal cells in each infection condition for in total $n = 1,129$ enterocyte cells

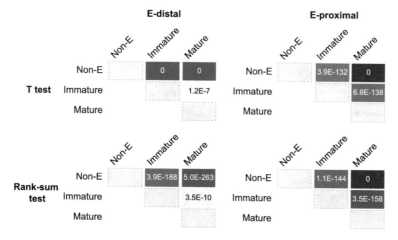

Fig. 2.18 Statistical difference (P-values from two-sided t-test and rank-sum test) of marker genes expression levels between immature versus mature, non-enterocyte (non-E) versus immature, non-E versus mature for E-distal and proximal clusters on mouse epithelial cell dataset

Interestingly, scScope further subdivided this cell type into four subpopulations: differential expression of markers provided delineation of distal vs. proximal enterocyte subpopulations, while expression levels of these markers provided further delineation into immature vs. mature subtypes. The assessment of maturity was based on expression levels of distal or proximal markers (Figs. 2.17b and 2.18). Identifying these refined enterocyte subpopulations allowed us to make predictions about specific cell-type response to infection. For example, the pro-inflammatory gene *Saa1* was overexpressed during both *Salmonella* and *H. polygyrus* (day 10) infections in distal enterocytes, but not in proximal enterocytes (Fig. 2.17c). This geographic pattern of *Saa1* expression is known for *Salmonella* infection, but is a novel prediction for *H. polygyrus* infection. Thus, scScope can be used to rapidly explore scRNA-seq data across perturbed conditions to predict novel gene function and identify new cell subtypes.

2.5.3   Biological Discovery of Mouse Brain Data

The 1.3M cells in the brain dataset were obtained from multiple brain regions, including the cortex, hippocampus and ventricular zones, of two embryonic mice. scScope automatically identified 36 clusters, and we assigned each cluster to one of three major cell types based on criteria from the Allen Brain Atlas (http://brain-map.org) (Fig. 2.19a): Glutamatergic neurons, GABAergic neurons and non-neuronal cells. he proportions of neurons and non-neurons identified by scScope were consistent with cell proportions reported by previous brain study (Fig. 2.19a). We investigated

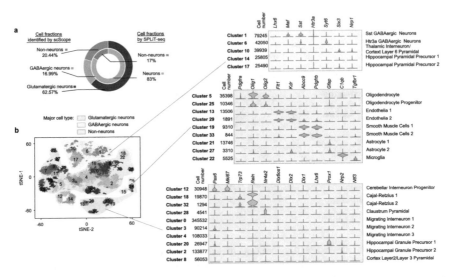

Fig. 2.19 Application of scScope to explore biology in 1.3M mouse brain dataset. **a** Fractions of three major cell types (glutamatergic neurons, GABAergic neurons and non-neurons) identified by scScope and comparisons with reported neuron fractions by previous SPLiT-seq research. **b** Left: scScope results visualized using tSNE on n = 30K cells (randomly sampled from the full dataset). Clusters were divided to three major types based on gene markers. Right: Large-scale annotation of clusters to known cell types according to top 10 overexpressed genes. Violin plots: expression distribution of marker genes for discovered clusters. Vertical axis (left): clusters with known cell type annotations and corresponding cell numbers. Horizontal axis: differentially expressed marker genes across shown clusters. Vertical axis (right): cluster annotation based on previously reported cell-subtype-specific genes

whether we could identify biological meaning to the 36 clusters, some of which contained fewer than 1,000 cells. Satisfyingly, by comparing our top overexpressed genes with known cell-type markers, we were able to assign two thirds of the clusters to known cell types (Fig. 2.19b). Thus, scScope is able to rapidly, automatically and directly identify bona fide, rare cell types from large and complex biological datasets.

2.6   Summary

The rapid development of sequencing technology and the increasing scale of datasets have posed severe challenges to existing analysis tools. This chapter introduces deep recurrent autoencoder, which provides a fast, accurate, and efficient method for large-scale single cell data analysis. At the same time, it can correct experimental batch errors, obtain low-dimensional data representation, and impute missing features and clustering. scScope offers a platform that will help keep pace with the rapid advances in scRNA-seq, enabling rapid exploration of heterogeneous biological states within large and noisy datasets of single-cell transcriptional profiles. The method proposed in this chapter is not based on the distribution assumption of genetic data, thus its application is not limited to the single-cell transcriptomics analysis.

Although the method in this paper solves the problem of cell type analysis on large-scale single-cell data, the model contains multiple hyperparameters, such as feature dimensions, recurrent iterations, and number of autoencoder layers. Using the same setting for different datasets is obviously unreasonable. Therefore, exploring the parameter setting of different data is a further optimization direction; at the same time, in the evaluation of the method, this chapter uses both simulation data and real data. For simulation data, the simulation model is not equivalent to the real sequencing process and there will be model deviations, which will cause the generated data to not be completely equivalent to the real data. For the real data, although the data is actually collected, the real cell types in the organization are unknown and only the conclusions of other studies can be used as a "partial" reference. How to set up an effective evaluation system is also an area worthy of further exploration. Single-cell transcript analysis is still a new and rapidly developing field and the method proposed in this chapter still needs to be further tested in actual data analysis applications.

References

1. Gawad C, Koh W, Quake SR (2016) Single-cell genome sequencing: current state of the science. Nat Rev Genet 17(3):175
2. Saliba AE, Westermann AJ, Gorski SA, Vogel J (2014) Single-cell rna-seq: advances and future challenges. Nucleic Acids Res 42(14):8845–8860
3. Shalek AK, Satija R, Shuga J, Trombetta JJ, Gennert D, Lu D, Chen P, Gertner RS, Gaublomme JT, Yosef N et al (2014) Single-cell rna-seq reveals dynamic paracrine control of cellular variation. Nature 510(7505):363
4. Macosko EZ, Basu A, Satija R, Nemesh J, Shekhar K, Goldman M, Tirosh I, Bialas AR, Kamitaki N, Martersteck EM et al (2015) Highly parallel genome-wide expression profiling of individual cells using nanoliter droplets. Cell 161(5):1202–1214
5. Zheng GXY, Jessica TM, Belgrader P, Ryvkin P, Bent ZW, Wilson R, Ziraldo SB, Wheeler TD, McDermott GP, Zhu J et al (2017) Massively parallel digital transcriptional profiling of single cells. Nat Commun 8:14049
6. Han X, Wang R, Zhou Y, Fei L, Sun H, Lai S, Saadatpour A, Zhou Z, Chen H, Ye F et al (2018) Mapping the mouse cell atlas by microwell-seq. Cell 172(5):1091–1107
7. Pierson E, Yau C (2015) Zifa: dimensionality reduction for zero-inflated single-cell gene expression analysis. Genome Biol 16(1):241
8. Risso D, Perraudeau F, Gribkova S, Dudoit S, Vert JP (2018) A general and flexible method for signal extraction from single-cell rna-seq data. Nat Commun 9(1):284
9. Wang B, Zhu J, Pierson E, Ramazzotti D, Batzoglou S (2017) Visualization and analysis of single-cell rna-seq data by kernel-based similarity learning. Nat Methods 14(4):414
10. Cleary B, Cong L, Cheung A, Lander ES, Regev A (2017) Efficient generation of transcriptomic profiles by random composite measurements. Cell 171(6):1424–1436
11. Van Dijk D, Sharma R, Nainys J, Yim K, Kathail P, Carr AJ, Burdziak C, Moon KR, Chaffer CL, Pattabiraman D et al (2018) Recovering gene interactions from single-cell data using data diffusion. Cell 174(3):716–729
12. Butler A, Hoffman P, Smibert P, Papalexi E, Satija R (2018) Integrating single-cell transcriptomic data across different conditions, technologies, and species. Nat Biotechnol 36(5):411

13. Lopez R, Regier J, Cole MB, Jordan MI, Yosef N (2018) Deep generative modeling for single-cell transcriptomics. Nat Methods 15(12):1053
14. Eraslan G, Simon LM, Mircea M, Mueller NS, Theis FJ (2019) Single-cell rna-seq denoising using a deep count autoencoder. Nat Commun 10(1):390
15. Levine JH, Simonds EF, Bendall SC, Davis KL, El-ad DA, Tadmor MD, Litvin O, Fienberg HG, Jager A, Zunder ER et al (2015) Data-driven phenotypic dissection of aml reveals progenitor-like cells that correlate with prognosis. Cell 162(1):184–197
16. Zappia L, Phipson B, Oshlack A (2017) Splatter: simulation of single-cell rna sequencing data. Genome Biol 18(1):174

Chapter 3
Computational Recovery of Information From Low-quality and Missing Labels

This chapter focuses on another type of missing data: the classification problem under missing data labels. Missing data labels is a common problem in the context of large-scale data analysis, which challenges traditional supervised learning methods. Especially in life science research, as the data labeling require a strong professional research background and expert experience, the lack of labels is an unavoidable problem for large-scale biological data. This chapter presents a robust information theoretic (RIT) model to reduce the uncertainties, i.e. missing and noisy labels, in general discriminative data representation tasks. The fundamental pursuit of our model is to simultaneously learn a transformation function and a discriminative classifier that maximize the mutual information of data and their labels in the latent space. In this general paradigm, we respectively discuss three types of the RIT implementations with linear subspace embedding, deep transformation and structured sparse learning. In practice, the RIT and deep RIT are exploited to solve the image categorization task whose performances will be verified on various benchmark datasets. The structured sparse RIT is further applied to a medical image analysis task for brain MRI segmentation that allows group-level feature selections on the brain tissues.

3.1 Introduction

Data transformation is perhaps the most prevalent and effective approach to be adopted when dealing with real-world image of high dimensionality. Transforming high-dimensional image into a latent space is plausible due to its two prominent advantages in data compression and feature learning. In this paper, we will focus on the discriminative data transformation approaches that incorporate labels into the learning phase. While this task-driven feature learning topic has been discussed in

© Tsinghua University Press 2021
F. Bao, *Computational Reconstruction of Missing Data in Biological Research*,
Springer Theses, https://doi.org/10.1007/978-981-16-3064-4_3

some previous works, there are three important issues that are hardly addressed, or at least not simultaneously considered.

First, in typical embedding framework, the representation and classification functions are trained sequentially. Such training procedures convey no classifier information into the feature learning part. It makes much sense if one can train the classifier and transformation simultaneously to encourage the most suitable features for a particular task. Secondly, for real world problems, the acquisition of labels are very expensive. In cases of insufficient labels, the discriminative learning results cannot capture the whole structures of the dataset. Thirdly, even though plenty of labeled data are available, in some cases, their labels are not definitely reliable. The noisy labels may potentially cause bias in both the features and the classifiers.

To address the aforementioned three challenges, in this paper, we propose a robust information theoretic embedding (RIT) algorithm by exploiting the mutual information as the discriminative criteria. Different from previous works, it simultaneously learns an transformation function and a probabilistic classifier to classify the points in the latent space. The incorporated probabilistic classifier, i.e. a multinomial logistic regression [1] (a.k.a. soft-max function), does not only encourage the class margins but also defines the probability density function (PDF). Such well-defined PDF facilitate the calculations of the conditional entropy in the latent space that is helpful to detect noisy labels.

The aforementioned RIT learning framework provides a general paradigm for feature learning. It seamlessly works with different data transformation functions to address diverse machine learning tasks. In this paper, as most subspace models, we first consider the most intuitive and basic implementation of RIT with the linear transformation. A toy illustration of this linear RIT model and its corresponding robust embedding results are provided in Fig. 3.1. Apparently, when there are unsupervised and noisy labeled samples, RIT significantly outperforms other methods from both the visualization effects and quantitative evaluations. In addition to this basic version, two other types of sophisticated data transforming strategies will be also considered in the RIT paradigm.

In the first extension, the deep learning (DL) concept [2] is incorporated into the RIT framework. Unlike linear subspace model, DL performs nonlinear transformation on the raw data by a deep neural network. The advantages of deep RIT are mainly concluded as two points. First, it adopts the deep structure to hierarchically transform information from layer to layer that is proven to be more effective than the shallow functions. Besides, deep RIT takes the advantage the information theoretic quantity as the learning objective, which could potentially reduce the label uncertainties in the training set. The performances of this deep RIT model will be verified on some image datasets, e.g. ImageNet [3] for image categorization.

In the second extension, we enhance the robustness of RIT by introducing the prevalent structured sparse norms into it. Structured sparse norm does not encourage entrylevel sparseness as conventional ℓ_1-type sparse problems. Instead, it enhances the sparsity on a group of variables. Such plausible mechanism allows RIT to select more reasonable feature groups in the subspace that sheds light on feature learning sides. In a nutshell, the structured sparse RIT (SS-RIT) exhibits two significant

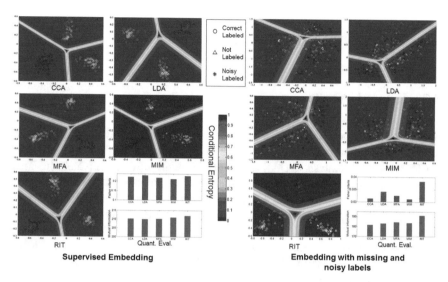

Fig. 3.1 The embedding results of the faces from three categories in Yale-B dataset. Different colors represent different classes. The left panel are the embedding results for data that are all correctly supervised and the right panel are embedding results with missing labels and incorrect labels. The last figure in each panel quantitatively evaluates the class separability in the embedding space by calculating both the mutual information and the Fisher criteria. For embedding results in the right panel, we use the semi-supervised version of the algorithm if it can be extended with a Laplacian regularization. The background colors represent the conditional entropy for each point in a 2D space. To calculate the conditional entropy for other embedding methods, we first project the data into the latent space and then fit a multinomial logistic regression in the latent space

advantages: (1) employing the information theoretic approaches to reduce the uncertainties among labels and (2) incorporating structured sparsity-induced norms for group-level feature learning. We will apply the SS-RIT to a challenging task of brain magnetic resonance image (MRI) segmentation where both label and feature uncertainty occur.

In sum, the contributions of this paper are mainly summarized as two-folds:

- We present an information theoretic learning framework that is able to conduct feature learning and classification jointly. The RIT model is robust to the uncertainties in the training data and could achieve reliable performances even though with missing and noisy labels.
- The proposed RIT is a flexible feature transformation framework that seamlessly works with different types of feature transformations, e.g. deep and structured learning, to cope with diverse practical problems.

The remaining of the paper is organized as follows: Section 3.2 reviews related works on discriminative feature learning. The proposed RIT learning framework is introduced in Sect. 3.3.1 and its detailed implementations with various feature learning functions are discussed in Sect. 3.3.4. The last evaluates the performances linear RIT, Deep RIT and sparse structured RIT on different tasks.

3.2 Related Works

Subspace models are widely used in the machine learning field for data representation. Statistic methods, such as Principle Component Analysis (PCA) [4, 5], Linear Discriminant Analysis (LDA), Canonical Correlation Analysis (CCA) [6, 7]are early attempts. Even after decades from their original propositions, the effectiveness of these methods are still not to be despised. Manifold learning [8, 9]and its variants [10, 11] find projections that optimally preserves the graph distances of high dimensional data. The graph structure enables the exploration of various nonlinear graph-based similarities, e.g. geodesic distance [12], commute time [11] and locality similarity [8], to describe the intrinsic relationship among data. Marginal Fisher Analysis (MFA) is the discriminative manifold learning method that extends Fisher discriminant criteria into a manifold [9]. Discriminative Locality Alignment (DLA) combines the locality similarity metric and the marginal sample weighting strategy. DKA more reasonably utilizes the local information of data and thus leads to more robust performance. Generalized Multiview Analysis (GMA) seeks an optimal subspace by solving a quadratic constrained quadratic program (QCQP) over different subspaces (e.g. PCA, LDA, MFA). The promises of these generalized models have been witnessed in a number of benchmark datasets.

Unlike subspace based method, dictionary-based models do not impose orthogonal restrictions on the projections, allowing more flexibility to adapt the representation to the data. Within the dictionary learning framework, some priors can be placed on the dictionary to encourage the desired data structure. Widely used priors include non-negative [13], sparse [14–16] and structured sparse [17]. In [18], discriminative dictionaries are generated by incorporating a discriminative function into the task-driven framework.

Deep learning is an emerging technique which has wide influences in the big-data industries [2, 19, 20]. It tackles a fundamental problem in machine learning: how to generate informative features in a task-driven manner. While the deep learning has been extensively used to reduce the noises in the raw data, less efforts have been devoted to handle the noises in the labels. In fact, label uncertainty is meanwhile a critical problem that needs careful considerations. Different from existing works, in this paper, we introduce a new learning objective into the typical DL framework from a novel perspective of information theory.

In the community of machine learning, the information theoretic quantities have been used for data representation [21], clustering [22] and feature selection [23, 24]. For data embedding, Mutual Information Maximization (MIM) has been proposed to extract features in a discriminative manner [25]. Its formulation resembles the Fisher discrimination but defines the discriminant via information quantity. Although both [25] and our RIT model share one common thing in utilizing mutual information as the discriminative criteria, the two algorithms are quite different. In MIM, all the probability density function are estimated via the nonparametric way. Therefore, MIM is a fully supervised embedding method and is sensitive to the quality of labels. RIT provides a more flexible way to interpret the probabilities with a probabilistic classifier which can be easily extended to semi-supervised version and is robust to noises in the given labels. Our RIT model was inspired by the RIM work in on information theoretic function design. However, RIM just considers discriminative clustering in the original data space without any feature learning mechanism involved. The major concern of RIT is about feature learning. It considers generating more reasonable feature representations to enhance the discriminative structure in a transformed space. In detail, we will consider three data transformation functions in this work including subspace, deep and structured sparse transformations.

3.3 Robust Information Theoretic Learning (RIT)

In this part, we will introduce the robust information theoretic embedding (RIT) model and its solutions.

3.3.1 Model

For a flexible description, we adopt a probabilistic framework to address the task of discriminative learning. In probability theory and information theory, the mutual information is a quantity that measures the mutual dependence of the two random variables. It measures how much knowing one of these variables reduces uncertainty about the other. Therefore, in our formulation, the mutual information serves as the basic discriminant criteria to measure the class separability in the transformed space.

For the ease of illustration, we define $\mathbf{x}_i \in \mathbb{R}^n$ as the original data obtained in real world and $\mathbf{y}_i = g(\mathbf{x}_i) \in \mathbb{R}^m$, $m < n$ as the corresponding point of $\mathbf{x}_i$ in the latent space. $g(\cdot)$ is a transformation. $l_i = k$ means that the i point belongs to the k class, $k = 1...C$. In Shannon's information theory, the mutual information of latent points and labels, i.e., $I(\mathbf{L}, \mathbf{Y})$, can be expressed in the following form,

$$I(\mathbf{L}, \mathbf{Y}) = H(\mathbf{L}) - H(\mathbf{L}|\mathbf{Y})$$
$$= -\int p(l) \log p(l)dl + \iint p(l, \mathbf{y}) \log p(l|\mathbf{y})d\mathbf{y}dl \qquad (3.1)$$

where $H(\cdot)$ denotes the entropy.

As shown in (3.1), the mutual information can be expanded as the summation of two entropy terms. The conditional entropy $H(\mathbf{L}|\mathbf{Y})$ reveals the total uncertainty of labels by observing the latent features. Therefore, it should be minimized. As indicated in [22, 26], this conditional entropy implicitly represents the margins between different classes. A small conditional entropy corresponds to a large margin. Besides, the entropy $H(\mathbf{L})$ encodes the label distribution which is always maximized in semi-supervised learning to avoid label bias on some specific classes.

The estimation of information theoretic quantities depends on probability density function (PDF) of transformed data and labels. In the RIT model, we assume that the data in the latent space can be well separated by a probabilistic classifier. In machine learning, one extensively used probabilistic classifier is the multinomial logistic regression (MNL). Without the loss of generality, in this paper, we exploit the MNL in the RIT formulation. We assume there are C classes in total and get C pairs of $\theta_j = (\mathbf{w}_j, b_j)$ in the parameter space of the MNL. It is worth noting here that l is not the supervised label. In fact, it is the label assigned by the MNL in the latent space. With the MNL, the conditional probability $p(l_i = k | \mathbf{y}_i)$ is explicitly defined,

$$p_{ik} = p(l_i = k | \mathbf{y}_i) = \frac{\exp(\mathbf{w}_k^T \mathbf{y}_i + b_k)}{\sum\limits_{j=1}^{C} \exp(\mathbf{w}_j^T \mathbf{y}_i + b_j)}. \tag{3.2}$$

According to (3.1), it is obvious that this kind of implicit labels will be integrated out in the calculation of the mutual information. Of course, there is another kind of supervised labels, i.e., l^s, which are explicitly provided by the user. Accordingly, we treat data as two kinds regarding whether their labels are explicitly given or not. For training, we assume there are N feature vectors in total, i.e. $\mathbf{Y} = \{\mathbf{y}_1, \mathbf{y}_2 ... \mathbf{y}_N\}$. Among these N data points, we get t supervised features, i.e. $\mathbf{Y}^s = \{\mathbf{y}_1^s, \mathbf{y}_2^s ... \mathbf{y}_t^s\} \subset \mathcal{S}$, with their labels explicitly provided as $\mathbf{L}^s = \{l_1^s, l_2^s ... l_t^s\}$.

We define $\mathbf{X} \in \mathbb{R}^{n \times N}$ are the original feature and $\mathbf{Y} \in \mathbb{R}^{m \times N}$, $m < n$ are the points in the latent space. $g(\cdot)$ is the mapping or data transformation function. In this part, we consider the most widely used linear transformation, i.e., $\mathbf{Y} = \Omega \mathbf{X}$, $\Omega \in \mathbb{R}^{m \times n}$. More general data transformations will be discussed in Sect. 3.3.4, Accordingly, we give the general form of the robust information theoretic embedding (RIT) model,

$$\min \ -I(\mathbf{L}, \mathbf{Y}) - \lambda \mathcal{C}(\mathbf{L}^s, \mathbf{Y}^s)$$
$$s.t. \ \mathbf{Y} = g(\mathbf{X}) \tag{3.3}$$

In the objective function of (3.3), the first term is the mutual information of all the data no matter whether they are supervised or not. The second term $\mathcal{C}(\mathbf{L}^s, \mathbf{Y}^s)$ is the regularization by penalizing the loss of the probabilistic classifier which only involves supervised data and their labels. We will discuss its expression and effectiveness in the next subsection in 3.5. Till now, why our RIT model naturally handles semi-supervised embedding tasks is selfevidently. It utilizes all the samples (both supervised or not) in the mutual information term and the supervised information are further penalized in the second term.

3.3.2 RIT with Noisy Labels

RIT is meanwhile very robust to the noises in the supervised labels. This desired advantage owes to two points. Firstly, the objective of RIT does not only overfit the losses of supervised data. In addition to the logit loss $\mathcal{C}(\cdot)$ in 3.3, RIT simultaneously seeks for a balance to maximize the mutual information term, which does not rely on the supervised label. It is conceivable that an overfitted logistic machine may achieve a good score on the loss of $\mathcal{C}(\mathbf{L}^s, \mathbf{Y}^s)$. However, such a bad logistic classifier learned from noisy labels may not achieve a good score on the mutual information. Therefore, the mutual information term alleviates the disturbance of the noisy labels.

Secondly, the MNL itself could also contribute to alleviating the noisy labels. In detail, $p(l_i^s|\mathbf{x}_i^s)$ exactly reveals the uncertainty of the supervised labels by observing the features. It is conceivable that a well trained MNL could not fit all the data perfectly. Particularly, it hardly fits the outliers in the training set. Therefore, the noisy labeled data generally exhibit small conditional probability implying that they cannot be well explained by the current MNL. Accordingly, following the idea in [27] we can define the weight $\phi_i = p(l_i^s|\mathbf{x}_i)$ for the ith supervised sample and incorporate this quantity to design a weighted MNL,

$$\mathcal{L}(\mathbf{L}^s, \mathbf{Y}^s) = \prod_{i=1}^{t} (p_{ik})^{\phi_i}, \qquad (3.4)$$

The function $\mathcal{L}$ is not a likelihood in the usual sense; but it has much general meaning to alleviate the disturbances of outliers in the training set. With the weighted likelihood, we get its log-likelihood expression and the cost function is subject to the following equation, *i.e.*

$$\mathcal{C}(\mathbf{L}^s, \mathbf{Y}^s) = \sum_{i=1}^{t} C_i = \sum_{i=1}^{t} \phi_i \log p_{ik}. \qquad (3.5)$$

From the log-likelihood, obviously, when ϕ_i is small, the ith sample contributes less to the global cost. C_i is the loss that the ith supervised sample contributed to the global objective. On the contrary, a large weight enhances the effectiveness of the ith sample to the optimization. Therefore, to make a robust embedding, it is plausible if we can denote small weights to the samples whose labels are wrongly supervised. Fortunately, within the probabilistic framework, it is possible to define such kind of weight by the conditional probability returned by the MNL.

The weight is updated along with the processing of the whole RIT optimizations. Till now, the cost $\mathcal{C}(\mathbf{L}^s, \mathbf{Y}^s)$ exactly corresponds to the general losses used in weighed logistic regression. In the optimization, these weights are dynamically updated and the whole optimization is cast to a sequence of reweighted programming. The details of the optimization and the weight updating procedures are provided in 3.2.

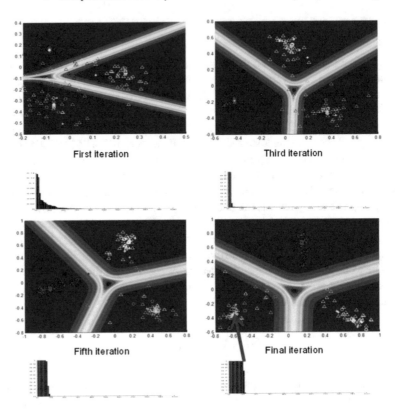

Fig. 3.2 The embedding results and the conditional probability for supervised data in different iterations of RIT optimization for the toy demo discussed in 3.1. In each subfigure, the histograms report $1 - p(l_i^s|\mathbf{x}_i^s)$, of each labeled point which are arranged in a decreasing order. The red bars indicate the noisy supervised data. Blue bars denote the samples whose labels are correctly supervised

From 3.2, it is obvious that with the processing of the iterations, the noisy labeled data are automatically identified by our algorithm (see the red bars). According to 3.4, it is apparent that these noisy data may contribute little to fit the logistic regression and their effectiveness to discrimination are only represented in the mutual information term that does not rely on the supervised labels. However, our algorithm cannot perfectly alleviate all the disturbances of noisy labels, it is found from the last subfigure in 3.2 that one noisy point (indicated by the arrow) is still embedded to a wrong place.

3.3.3 RIT Subspace Model

We show the optimization of the RIT model in this part. With p_{ik} defined in (3.2) and $p_k = \frac{1}{N} \sum_{i=1}^{N} p_{ik}$, we give the empirically estimation of the mutual information term that,

$$I(\mathbf{L}, \mathbf{Y}) = \frac{1}{N} \sum_{i=1}^{N} I_i = \frac{1}{N} \sum_{i=1}^{N} \sum_{k=1}^{C} \{p_{ik}[\log p_k - \log p_{ik}]\} \tag{3.6}$$

The term I_i defines the loss that the ith sample contributed to the mutual information [22]. For term $C(\cdot)$, we choose it to be the log-weighed-likelihood of the losses of MNL. As discussed previously, the optimization involves two variables, *i.e.* the data transformation Ω and the MNL parameters $\theta_i = (\mathbf{w}_i, b_i), i = 1..C$.

The gradient of the two terms with respect to Ω, $\mathbf{w}$ and b can all be derived through chain rule. For example, $\frac{\partial I}{\partial \Omega} = \frac{1}{N} \sum_{i=1}^{N} \sum_{k=1}^{C} [\log \frac{p_k}{p_{ik}} + 1]\frac{\partial p_{ik}}{\partial \Omega}$, $\frac{\partial C}{\partial \Omega} = \frac{1}{N} \sum_{i=1}^{N} \frac{1}{p_{il_i^s}} \frac{\partial p_{il_i^s}}{\partial \Omega}$. This chain rule is also applied to the derivatives for $\mathbf{w}$ and b. The only modification is to change the partial derivative of Ω to be the partial derivatives with $\mathbf{w}$ and b, repectively.

After getting the derivatives, the whole RIT optimization is solved in an alternating framework. We denote the objective in (3.3) as $f(\mathbf{X}, \mathbf{L}^s | \Omega, \theta, \Phi)$, where Φ is the weight matrix. The updating rule is provided by $\theta^{k+1} = \arg\min f(\mathbf{X}, \mathbf{L}^s | \Omega^k, \theta, \Phi^k)$ and $\Omega^{k+1} = \arg\min f(\mathbf{X}, \mathbf{L}^s | \Omega, \theta^k, \Phi^k)$. Both the updating of Ω and θ depend on the gradient descent method and we use the LBFGS quasi-Newton optimization algorithm[1] to get a fast and robust convergence.

3.3.4 RIT Extensions

In the previous part, the general paradigm of RIT subspace model with linear transformation has been discussed. In fact, RIT is a robust information theoretic feature learning framework that works friendly with many kinds of data transformation functions. As extensions, we will introduce other two types of prevalent data transformation strategies into the RIT framework from the perspectives of deep learning and structured sparse learning.

Deep RIT

In this part, we show how to incorporate deep learning concepts into RIT to improve the performances of feature learning. A schematic summarization of the deep RIT (DRIT) model has been provided in 3.3 which is mainly composed of two

---

[1] http://www.cs.ubc.ca/~schmidtm/Software/minFunc.html.

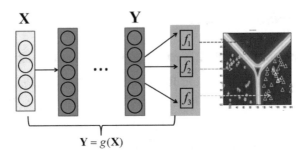

Fig. 3.3 The schematic illustration of the deep RIT (DRIT) model by exploiting deep neural network as the data transformation function $g(\cdot)$. In the classification layer (green layer) of the DNN, the classifier assigns different points to its corresponding category by maximizing the information theoretic term in 3.3 as the learning objective

parts of feature transformation (blue layers) and task-driven learning parts (green layer).

At a glance, the deep neural network (DNN) plays the role of a mapping function $g(\cdot)$ in 3.3, that transforms the input data/image (yellow layer $\mathbf{X}$) into a high-level representation $\mathbf{Y}$. We follow the general principle in the field to define the activations of the neural network. In details, the jth node of the $(l)^{th}$ layer is connected to the nodes on the $(l-1)^{th}$ layer with parameters $\theta^{(lj)} = \{\mathbf{w}^{(lj)}, \mathbf{b}^{(lj)}\}$ $i.e.$

$$o^{(lj)} = \gamma(a^{(lj)}),\, a^{(lj)} = \mathbf{w}^{(lj)}\mathbf{o}^{(l-1)} + b^{(lj)}, \tag{3.7}$$

where $\gamma(\cdot)$ is a nonlinear mapping; $\mathbf{w}^{(lj)}$ and $b^{(lj)}$ are the weights and bias. The number of hidden layers of this deep representation part can be large. In the image analysis tasks, the dimensions of input data are very high and the convolutional operations are always used in the first coupe of layers of the DNN. In this part, we do not prefer to specify the detailed DNN structure because DRIT works well with many types of deep configurations. The specific DNN setting that is used in this work for the image categorization will be discussed in the corresponding experimental part.

We remark on the differences between DNN transformation and the linear transformation in the RIT subspace model. In the previous linear model, only a mapping matrix Ω is optimized while this deep transformation involves millions of hidden parameters. Such a large amount of parameters allow hierarchical transformations that could potentially increase the chances to get a better representation $\mathbf{Y}$ on the last representation layer. On the other hand, deep training also imposes great computational burdens and requires tons of training samples.

After obtaining the latent representation $\mathbf{Y}$ of the DNN, a multinomial logistic regression layer (green layer) is connected to it for data classification. In 3.3, for simplicity, only three classes and their corresponding classifiers $f_1, \ldots, f_3$ are shown. In typical DNN training, the objective function always directly minimizes the logistic loss to make minimal prediction error on the training set. In the DRIT model, we further utilize the information theoretic quantity defined in (3.3) to reduce the uncertainties in the supervised labels. Such objective generalizes all the nice properties of the RIT learning to this deep learning framework. More importantly, similar as the methods in Sect. 3.3.2, DRIT also exhibits the plausible mechanism to detect wrongly supervised labels in the training set which will be experimentally discussed later.

While DRIT exploits the RIT as the learning objective, typical back-propagation (BP) algorithm still easily applies to solve it. To note, we have denoted two terms, *i.e.* $C(\cdot)$ and $I(\cdot)$ in the RIT objective. These two terms should be simultaneously considered in the BP process to adjust parameters in DNN. In general, the gradient for the parameter in DNN can be determined by the following additive formulation.

$$
\frac{\partial C}{\partial \theta^{(lj)}} = \sum_i \delta(i \in S) \underbrace{\left(\frac{\partial C_i}{\partial o_i^{(lj)}}\right) \frac{\partial o_i^{(lj)}}{\partial a_i^{(lj)}} \frac{\partial a_i^{(lj)}}{\partial \theta^{(lj)}}}_{BP}
$$
$$
+ \sum_i \underbrace{\left(\frac{\partial I_i}{\partial o_i^{(lj)}}\right) \frac{\partial o_i^{(lj)}}{\partial a_i^{(lj)}} \frac{\partial a_i^{(lj)}}{\partial \theta^{(lj)}}}_{BP}. \tag{3.8}
$$

The above gradient for parameter $\theta^{(lj)}$ is easily verified according to the chain rule. The terms in the brackets come from error back propagation (BP) and the remaining terms out of the brackets are easily calculated with the matrixform derivations. In the above formulation, we have used S to denote the set of supervised data points and $\delta(\cdot)$ is the indicator function which is 1 if $i \in S$ and 0 otherwise. I_i and C_i have been defined in (3.1) and (3.5), respectively. In practical training, we adopt the stochastic gradient descent strategy to update the parameters with parallel computing [28].

Structured Sparse RIT

In many practical problems, the data themselves exhibit certain structure. In this part, we will show how to effectively exploits this feature structure information to improve the performance of RIT. From a general view, the linear RIT model is comparable with a regression problem with each dimension in the latent space sharing its own regression vector. To support this claim, we recall the linear transformation in the RIT model that for any point $\mathbf{x}_i \in \mathbb{R}^n$, there exists a mapping $\mathbf{y}_i = \Omega \mathbf{x}_i$. We know that $\mathbf{y}_i = [y_{i1}, y_{i2}...y_{im}] \in \mathbb{R}^m$. Consequently, it is straightforward to get a regression-type formulation that $y_{iu} = \omega_u \mathbf{x}_i, \forall i = 1...N$ where $\omega_u \in \mathbb{R}^{1 \times n}$ is the uth row of Ω, *i.e.* $\Omega = [\omega_1^T ...\omega_u^T ...\omega_m^T]^T$.

From the discussions aforementioned, it is apparent that all the attributes in the feature vector $\mathbf{x}$ will contribute to the final embedding. In statistic, it has been widely investigated that such a dense regression is not the optimal one in most cases. In machine learning, one prevalent approach is to place sparse priors on the regression parameter to further improve the prediction accuracy of the model. Intuitively, sparse learning assumes that only a portion of factors in the original feature vector contribute to the learning process. The incorporation of the sparse norms facilitate automatic feature selection and alleviate the over-fitting problem for training data to a large extend.

However, in many cases, only constraining the sparseness of the factors does not seem appropriate because the considered factors are not only expected to be sparse but also to have a certain structure [17]. Therefore, structured sparsityinducing norms are now drawing more and more attentions in the community of machine learning, e.g. in grouped lasso. Therefore, it is nontrivial to incorporate structured regularization into the RIT model for the sake of better data interpretation.

Before introducing structured sparse RIT (SS-RIT). we will first introduce some sparse norms that play very critical roles in sparse learning. First, we define the general ℓ_p norm for a vector $\mathbf{a} \in \mathbb{R}^r$ as $\|\mathbf{a}\|_p = (\sum_i |a_i|^p)^{\frac{1}{p}}$. $\|\mathbf{a}\|_0$ denotes the ℓ_0 norm that counts the number of non-zero elements in $\mathbf{a}$. However, ℓ_0 norm is discrete and is analytically intractable. Therefore, its convex envelope, *i.e.* ℓ_1 norm is extensively used as a convex surrogate for sparse learning [29]. Based on the ℓ_1 norm, we will introduce the ℓ_1/ℓ_2 norm for structured sparse learning.

For the ease of explanation, we divide r dimensions of $\mathbf{a}$ into $|\mathcal{G}|$ overlapping groups, *i.e.* $G = 1, ... |\mathcal{G}|$, which implies that one attribute a_k can be assigned to different groups. We define $d_j^G > 0$ as the weight for the j variable in the G group. $d_j^G = 0$ means that the j attribute is excluded from the G group. Accordingly, the structured sparsity inducing norm can be defined as,

$$\Psi_{\mathcal{G}}(\mathbf{a}) = \sum_{G \in \mathcal{G}} \left\{ \sum_{j \in G} (d_j^G)^2 |a_j|^2 \right\} = \sum_{G \in \mathcal{G}} \|d_G \circ \mathbf{a}\|_2, \tag{3.9}$$

The operator $\circ$ is the component-wise product. The norm in (3.9) is called ℓ_1/ℓ_2 norm, because it encourages sparse selections at the group level and, in each group, the variables are densely penalized by a ℓ_2 norm.

Moreover, as indicated in [17], ℓ_1 norm is a specific case of (3.9) when $\mathcal{G}$ is the set of all singletons and with all the weights setting to 1. Accordingly, we present the general formulation of the structured sparse RIT model (SS-RIT) in (3.10) and the sparse RIT with ℓ_1 regularization is only a specific case of SS-RIT.

$$\min f(\mathbf{X}, \mathbf{L}^s | \Omega, \theta) + \mu \sum_{u=1}^{m} \Psi_{\mathcal{G}}(\omega_u) \tag{3.10}$$

In the above formulation, the first term is the loss of the RIT model in (3.3) and the second term is the structured sparse regularization of the projecting vector. ω_u is the uth row of the linear transformation Ω. With such a norm, it is apparent that each dimension y_u in the latent space is only associated with a number of attributes of $\mathbf{x}$ in the selected groups.

The optimization of SS-RIT is almost the same as the the solutions to the RIT model which depends on the alternation between Ω and θ. For the logistic parameters, they are irrelative to the added structured norms and thus the updating rule for RIT still applies to it without any change. However, in SS-RIT, the gradient of Ω now involves an extra term, *i.e.* $\Psi_G(\omega_u)$. To handle this structured sparsity-inducing norm, we follow the work in [17, 30] to introduce the variational variable η and solve the SS-RIT in a reweighted manner.

As following the result in [30], we have the following lemma,

Lemma 1 *For any matrix* $\mathbf{x}$*, its* ℓ_1 *norm is equivalent to the following problem with a variational variable* $\mathbf{z}$,

$$2 \parallel \mathbf{x} \parallel_1 = \min_{\mathbf{z} \in \mathbb{R}^p} \sum_{z=1}^{p} \frac{x_j^2}{z_j} + \parallel \mathbf{z} \parallel_1, \tag{3.11}$$

whose minimum is uniquely obtained for $z_j = |x_j|$.

Following (3.11), by defining η_u^G as the variable, $2 \sum_{u=1}^{m} \Psi_G(\omega_u)$, can be reformulated into the following variational form,

$$2 \sum_{u=1}^{m} \Psi(\omega_u) = \min_{(\eta_u^G)_{G \in \mathcal{G}}} \sum_{u=1}^{m} [\parallel (\eta_u^G)_{G \in \mathcal{G}} \parallel_1 + \sum_{G \in \mathcal{G}} \parallel \omega_u \circ d^G \parallel_2^2 (\eta_u^G)^{-1}] \tag{3.12}$$

By merging the variational variable and the ℓ_2 norm, Eq. (3.12) can be written in turn as $\min_{(\eta_u^G)_{G \in \mathcal{G}}} \sum_{u=1}^{m} \omega_u Diag(\zeta) \omega_u^T + \parallel (\eta_u^G)_{G \in \mathcal{G}} \parallel_1$ where the jth element in vector ζ is $\zeta_j^t = \sum_{G \in \mathcal{G}} (d_j^G)^2 (\eta_u^G)^{-1}$, $j = 1...n$. $Diag(\cdot)$ is an operation to write a vector into a diagonal matrix. Till now, the SS-RIT optimization in (3.10) SS-RIT is subject to the following variational optimization,

$$\min f(\mathbf{X}, \mathbf{L}^s | \Omega, \theta) + \frac{\mu}{2} \left\{ \sum_{u=1}^{m} \omega_u Diag(\zeta^t) \omega_u^T \right.$$
$$\left. + \parallel (\eta_u^G)_{G \in \mathcal{G}} \parallel_1 \right\}. \tag{3.13}$$

The problem in (3.13) is not jointly convex and led themselves well to simple alternating optimization scheme between $\Omega, (\eta_u^G)_{G \in \mathcal{G}}$ and θ.

The updating rules of Ω and θ is trivial following the gradient descend method. The updating rules of of the variational variable $\{(\eta_u^G)_{G \in \mathcal{G}}\}$ is given in lemma 1. In practice, $\{(\eta_u^G)_{G \in \mathcal{G}}\}$ is provided by

$$\{(\eta_u^G)_{G \in \mathcal{G}}\}^{k+1} \leftarrow \max\{\parallel \omega_u^k \circ d^G \parallel_2, \epsilon\}, \tag{3.14}$$

where $\epsilon \ll 1$ to avoiding numerical instability near zero. $\omega_u^k \in \Omega^k$ is the uth row of the optimal Ω^k obtained in the kth iteration. Till now, the whole SS-RIT optimization can be solved following the steps of alternating optimization. The whole optimization is regarded as converged when $\frac{\|\Omega^{k+1}-\Omega^k\|_F^2}{\|\Omega^k\|_F^2} < 10^{-3}$.

3.4 RIT for Image Classification

3.4.1 RIT Subspace Model

In this part, we investigate the performances of RIT on three benchmark image datasets including Yale-B face dataset [31], fifteen-scene dataset (Scene-15) and COIL-100 dataset [32].

In Yale-B face dataset, we simply use the cropped images in [31, 33] and resize them to 32×32 pixels. This dataset now has 38 individuals and around 64 near frontal images under different illuminations per individual. Fifteen scene dataset contains images from fifteen categories including both indoor and outdoor pictures. The COIL dataset contains the images of 100 objects from multi-views.

In the Yale-B dataset, we use the gray-scale pixel values on the raw face images to generate the feature vector. For the scene and object dataset, we follow the bag-of-feature method to extract visual features. In a nutshell, to describe an image, we use a grid-based method to extract the dense SIFT features. The dense SIIF features are extracted on 16×16 pixel patches sampled every 8 pixels. To generate features for fifteen scene and COIL dataset, the local sift features are assigned to a codebook with 1024 codewords by the kernel assignment [34] and lead to a final feature vector of $\mathbb{R}^{1024}$. For multi-view models, we also considered the gist feature [35] as another view.

For comparison purpose, we pit RIT against many prevalent subspace models including statistic methods (*e.g.* PCA [5], LDA[4], CCA [6]), DLA and GMA with LDA and MFA (termed as GMLDA and GMMFA), graph-based methods (*e.g.* MFA [9]), information theoretic learning [25] (MIM) and task driven sparse coding (TSC) with logistic regression as the objective [18]. For the ease of computational efficiency, before discriminative embedding, the original large feature vectors are pre-processed by PCA to a low dimensional subspace where 90% energy are preserved. In the implementation of RIT model, we fit $\lambda = 0.1$, and the learning procedures are regarded as converged when the changes of the objective is less than 10^{-4}. For multiple class categorization task, we follow the idea in [18] to train the model with the one-versus-all strategy. The experimental validations are divided into three parts, *i.e.* supervised embedding, semi-supervised embedding and embedding with noisy labels.

In the first test, we investigate the performance of RIT in a definitely supervised fashion. For training purpose, 30 samples per class in Yale-B dataset, 100 samples in each class of fifteen scene dataset and 40 images in COIL dataset are randomly selected as training samples. The rest images are used for testing and the experiments are repeated for 10 times. After data embedding, for data classification, we test three classifiers including Nearest Neighbors Classifier, Support Vector Machine (SVM)

Table 3.1 The classification results on different datasets with different data embedding algorithms and clasifiers-1 (Accuracy ± STD %)

Methods	Nearest neighbor classifier			SVM		
	Yale-B	Scene-15	COIL-100	Yale-B	Scene-15	COIL-100
Raw	67.6±2.8	55.1±2.4	62.2±2.1	89.6±2.1	75.1±1.5	81.7±2.2
PCA	78.6±1.7	62.1±2.6	73.2±1.5	88.5±1.7	71.5±2.5	80.6±1.3
LDA	89.6±1.3	65.6±2.1	75.7±1.6	92.7±1.2	74.2±2.1	82.9±1.4
CCA	90.3±1.6	66.4±2.3	73.8±1.7	92.3±1.3	75.9±2.2	83.7±1.5
MFA	91.7±1.4	68.2±2.2	74.1±1.5	92.9±1.3	76.8±2.3	83.3±1.6
MIM	90.7±1.7	66.1±2.7	76.1±1.5	93.3±1.6	75.6±2.4	83.9±1.7
TSC	89.9±1.5	67.2±2.4	77.6±1.7	94.5±1.2	76.5±2.5	85.3±1.4
DLA	91.2±1.3	67.3±1.9	75.1±1.3	93.7±1.2	76.7±2.4	85.7±1.7
GMLDA	89.2±1.1	68.7±2.1	78.1±1.9	92.1±1.1	76.6±2.3	86.7±2.1
GMMFA	89.5±1.1	67.9±2.3	77.9±2.3	91.8±1.4	77.3±2.4	87.1±2.1
RIT	91.3±1.6	68.9±2.6	77.2±1.5	93.9±1.3	77.3±2.3	86.2±1.5

and logistic regression. The best classification results of data embedding methods with different classifiers are reported in Tables 3.1 and 3.2. In Tables 3.1 and 3.2, the first row reports the classification accuracies on raw data as a comparison baseline. It is interesting to note, although less data dimensions are used in the latent space, the classification accuracies are even improved. This improvement owes to feature learning mechanism of discriminative data embedding [6, 25].

We consider the performances of RIT model to conduct semi-supervised discriminative embedding. As stated in Sect. 3.3, that resembles LDA but places a Laplacian term to encourage unlabeled points staying very close to the similar labeled points. The MIM model is a fully supervised model and we propose two possible ways to extend MIM to a semi-supervised version. First, we can initialize nonconvex MIM with the optimal projections learned by SDA (SDA+MIM). Besides, it is also possible to extend MIM to a semi-supervised version by incorporating a Laplacian term (LapMIM). Task driven sparse coding (TSC) [18] is straight-forward to be extended to the semi-supervised by using the labeled data in the classifier and keeping all the unlabeled data in the reconstruction term. DLA can also incorporate the unlabeled samples in the alignment stage and lead to the semi-supervised DLA (SDLA).

The results of semi-supervised learning results are reported in Table 3.3 by using the same feature and training samples as in Table 3.1. The classifier used in this test is the Logistic Regression. From the results, it is interesting to find the performances of discriminative embedding are further improved with some unlabeled points. By comparing the results with the supervised embedding results in Table 3.1, it is noted that semi-supervised-based embedding results exhibit smaller standard deviation.

From the experiments presented above, we find that among all the semi-supervised embedding methods, RIT achieves the best performances. According to previous discussions, other semi-supervised methods generally utilize a Laplacian term to

Table 3.2 The classification results on different datasets with different data embedding algorithms and clasifiers-2 (Accuracy ± STD %)

Methods	Logistic regression		
	Yale-B	Scene-15	COIL-100
Raw	88.7±2.6	74.6±2.3	83.4±1.9
PCA	87.3±1.6	70.6±2.4	81.1±1.5
LDA	92.2±1.3	75.3±2.1	83.1±1.3
CCA	92.7±1.4	76.1±2.2	83.8±1.4
MFA	93.1±1.5	76.3±2.2	83.3±1.5
MIM	92.5±1.7	76.7±2.3	82.7±1.8
TSC	93.8±1.3	77.8±2.4	85.8±1.6
DLA	93.7±1.3	78.3±2.5	84.2±1.5
GMLDA	92.7±1.2	78.3±2.4	88.2±1.7
GMMFA	93.4±1.3	78.7±2.6	87.2±1.4
RIT	94.3±1.2	79.8±2.4	87.3±1.3

Table 3.3 The classification results of semi-supervised discriminative embedding (Accuracy ± STD%)

Methods	Yale-B	Fifteen scenen	COIL-100
SDA	94.6±1.3	77.3±1.7	84.2±1.4
MIM+SDA	93.8±1.0	77.7±1.5	84.3±1.0
LapMIM	94.2±1.1	78.8±1.4	87.6±1.1
TSC	94.1±0.6	78.3±1.0	86.5±0.7
SDLA	95.5±1.2	80.2±0.9	88.2±1.2
RIT	96.3±0.9	81.6±1.3	89.7±0.9

regularize the unlabeled samples which shed no light on the discriminative side. RIT model directly enhances the discrimination of unsupervised points by optimizing the mutual information. Moreover, another significant advantage of RIT model is its flexibility in handling both supervised and semi-supervised embedding tasks. Other discriminative embedding models almost need extra modifications, e.g. adding another term into the objective, to adjust themselves to the semi-supervised version. Fortunately, RIT does not require any modifications in the model which is only determined by the training data type (supervised or semi-supervised) fed to it.

Consequently, we further consider a very challenging task that noisy labels are involved in the discriminative learning. To conduct the experiments, we randomly select a number of samples from training set and their labels are wrongly denoted. For each noisy level, the experiments are repeated for 10 times and the average classification accuracy on different datasets are reported in Fig. 3.4. We compare the RIT model with other benchmark data embedding and representation methods with the same training samples and noisy labels.

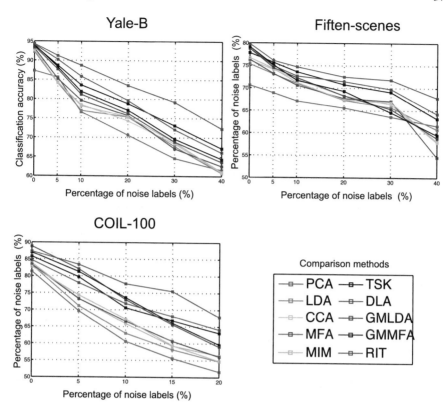

Fig. 3.4 The classification accuracy of different data embedding method with noisy labels

From the results, obviously, the performances of different data embedding methods gradually drop along with the noisy labels rate increasing. Fortunately, our RIT model is the most stable one to the noisy labels. Meanwhile, TSC achieves relatively good performances on this test. This is because TSC does not only address the discriminative objective and meanwhile considers optimal signal reconstruction. GMA also attains comparable results when the noisy label rate is small. Its performance further decreases along with the increases of noisy label number. For other discriminative embedding methods, *e.g.* LDA,CCA and MIM, their performances drop significantly with the increases of noisy labels. From the aforementioned discussions, we know RIT is the most robust one when compared to other discriminative embedding methods.

3.4.2 Deep RIT for Image Categorization

In this part, we evaluate the performances of DRIT model on three datasets. The first two are the fifteen scene and COIL-100 datasets which have been introduced and discussed in the previous part. In addition, a large-scale dataset, *i.e.* ImageNet [3] will also be used here. ImageNet task requires categorizing more than 100k testing images into 1000 classes which is quite challenging.

We choose the famous convolutional deep neural network (CDNN) proposed in as deep learning part because it has been widely regarded as the benchmark configuration in the field. In details, CDNN is composed of eight hidden layers: 5 convolutional layers (with pooling and ReLU nonlinear transformation) and 3 fully connected layers. We sought to the famous deep learning package Caffe [28] to realize both CDNN and DRIT. CDNN and DRIT share the same DNN structure in the feature transformation part but are trained with different learning objectives. CDNN directly use the logistic regression as the final layer while the proposed DRIT makes use of RIT function as the objective function.

We follow a standard protocol [28] to train the DNN by normalizing all the input images into the size of 224×224. The mean value on each RGB channel is subtracted from the original image. The DNN here involves more than 60 million hidden parameters to be learned from the data that requires a huge amount of training data. ImageNet dataset provides a sufficient amount of 1.2 million training images for deep learning. However, the fifteen scene and COIL datasets only contain limited number of training samples. Accordingly, on these two datasets, we follow the same idea in [47] to use the ImageNet training results to initialize the DNN. Then, the training images from these two datasets are used to fine-tune the parameters via back propagation. In practice, a random set of 100 and 40 images in each class of fifteen scene and COIL datasets are selected as the training samples. Both the average accuracy and standard deviations on these two datasets are reported in Table 3.4 ImageNet provides the training/testing list and only the accuracy on it is recorded.

We report the image categorization results of different deep learning methods in Table 3.4. We further consider the DNN with Hinge loss (HDNN) in the task layer. Hinge loss is also known as max margin loss which explicitly penalizes the margins of different classes. In addition to the categorization results with the logistic regression,

Table 3.4 The image categorization accuracy via deep learning

	Fifteen scene	COIL-100	ImageNet
HDNN	82.6±1.0	88.2±0.8	56.8
CDNN	82.3± 0.8	88.6±0.7	57.1
DRIT	83.4± 0.7	89.7±0.5	58.3
CDNN+SVM	82.9± 1.0	89.1±0.9	57.2
HDNN+SVM	83.2±1.1	87.7±0.9	57.3
DRIT+SVM	84.1± 1.1	90.4±0.9	58.6

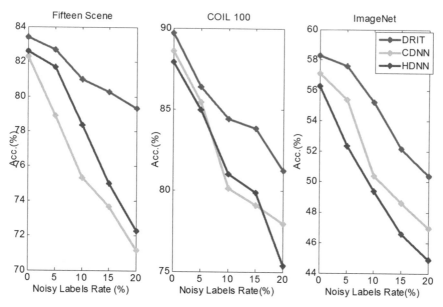

Fig. 3.5 The categorization accuracy of deep learning methods with different noisy label rates

the SVM classifier is also tested here. For SVM method, the values on the last representation hidden layer (before the categorization layer) are used to train a linear SVM classifier. From the results, it is apparent that deep learning strategy significantly outperforms the linear methods in Table 3.1. We have also tried the bag-of-feature methods (with 1000 codewords) on the ImageNet dataset and then classify them with linear RIT. Unfortunately, the accuracy is only around 26% which is far away from the deep learning results implying only deep learning could make reasonable predictions on this challenging ImageNet task. This is reasonable because DL is built upon millions of parameters while linear subspace model is only configured in a shallow framework. When comparing different deep learning results in Table 3.4, the advantages of DRIT model are self-evident. DRIT improves the accuracy for 1.2 points than typical CDNN on the ImageNet dataset. The improvements are also verified from the results with the SVM classifier where DRIT wins CDNN for 1.4% on ImageNet. The similar experimental findings are also made on the other two datasets.

The previous experiments on deep learning verify that the RIT is a better objective than logistic and hinge losses for general deep learning tasks. The advantages of RIT can be further highlighted on its robustness in reducing label ambiguity as discussed in Sect. 3.3.2. In this part, we further investigate DRIT's performances in treating label noises. We randomly select $p\%$ of training samples and denote definitely wrong labels to them. These wrongly supervised samples are mixed with ground truth training samples to conduct DNN learning. The noisy label rates are varied from 5% to 20% and, at each noisy rate level, the experiments are repeated for 10 times with the average accuracies reported in Fig 3.5. From the deep learning investigations in Table 3.4, it is found that SVM classifier achieves similar performances as logistic regression. Accordingly, in this part, only the deep learning results with logistic classifier are reported. By analyzing the results, we have observed the curves of DRIT suffer less drop than CDNN and HDNN. The results suggest that DRIT could cope with the noises in the training set much better leading to a relatively reliable curve with increases of the noisy label rates. The mechanism why RIT could alleviate this kind of noisy labels has been discussed in Sect. 3.3.2.

Till now, two implementations of RIT with linear transformation and deep transformation have been discussed and verified. From experimental comparison, it is concluded that deep RIT could always achieve much higher classification accuracy than the linear RIT model. However, the nature of DNN training requires sufficient training samples and heavy computational complexity. Therefore, for some challenging tasks, *e.g.* ImageNet, DRIT is strongly recommended due to its advancements in performances. On some small-scale dataset, *e.g.* COIL-100, the flexible linear RIT itself could already achieve very sound performances. Meanwhile, we have noticed that RIT and DRIT both exhibits the robustness in treating noisy labels in the training set. Both of them performs more reliable than other linear and deep approaches with noisy labels.

3.5 Biological Applications: SS-RIT for Brain MRI Segmentation

The discussions in the last section successful verify that RIT model is robust to label uncertainty. In many practical image analysis tasks, feature uncertainty is another critical issue that should be comprehensively considered. In this part, the discussions are extended on a practical task, *i.e.* MRI segmentation, where both the label and feature uncertainties simultaneously happen. In general, MRI segmentation is an important clinical task [36], that requires assigning brain tissues into white matter (WM), gray matter (GM) and cerebrospinal fluid (CSF) [36].

By analyzing the problem, the challenges mainly stem from two effects, *i.e.* partial volume and bias field effects [37]. In [38], it is revealed that partial volume effect and bias filed effect may respectively lead to the label uncertainty and feature uncertainty. The label uncertainty issue can be potentially well addressed by the RIT model whose robustness in conquering label ambiguity has been verified in the previous part. The remaining challenge here is to overcome the feature uncertainty. To cope with the bias effect, in the paper, we introduce a new strategy to conducting MRI segmentation on the super-voxel level. In details, the homogenous local regions on the brain tissue are grouped together into a super-voxel according to the method in [39]. An instance of super-voxels-level segmentation generated by the SLIC algorithm [39] has been provided in Fig 3.6.

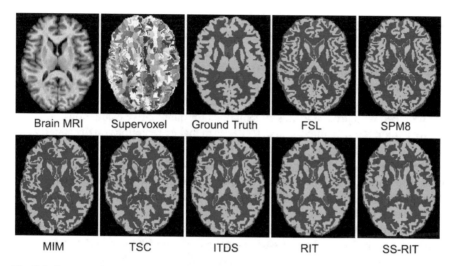

Fig. 3.6 Segmentation results of one brain MRI from the IBSR dataset

Then, on each super-voxel, multiple feature descriptors can be generated from different views that provides comprehensive quantitative summarizations of the coherent structure on the tissue. In this paper, we divide the visual descriptors into four groups as intensity, texture, SIFT and HOG features. The intensity feature is extracted by computing the intensity histogram with 64 bin. Local binary pattern [40] is exploited as a texture descriptor and can be summarized in a 36-dimensional feature vector. SIFT [41] is calculated on each super-voxel and leads to a vector in 128 dimensions. Finally, 31-dimension histogram of oriented gradient [42] is extracted. In total, a 259 dimensional feature vector is generated for each supervoxel.

The feature extraction strategy on super-voxel depicts the tissue content by descriptors from multiple views, avoiding the description biases from a single view. However, a natural question consequently raised here: which type of visual descriptors and their combinations are most suitable to the brain segmentation tasks? The structured sparse norm discussed in the above Section well solves this issue by enabling group-level feature selection/combination in subspace. Different from typical ℓ_1-norm-based sparse feature learning, structured sparse norm pays particular attentions to the physical structure of each feature group. It encourages the sparsity only at the group level avoiding destroying the original structure in a feature group. It thus makes more reasonable high-level summarizations of the original data [17] by keeping their inherent information content.

In summary, SS-RIT is a plausible paradigm in coping with the challenging brain MRI segmentation. First, the label uncertainty from partial volume effects is solved by the RIT model. Moreover, after combining the structured sparse norm into RIT, the SS-RIT naturally exhibits the group-level feature learning mechanism when generating the projection matrix. Finally, SS-RIT simultaneously performs data embedding and classification in the joint framework, which is flexible and effective in enhancing the discriminative structure of the data points in the latent space.

The experiments were conducted on two widely used datasets from Internet Brain Segmentation Repository (IBSR) [43] and BrainWeb database [44]. The IBSR dataset consists 18 real images with a size of $256 \times 256 \times 128$ voxels. BrainWeb dataset consists of 18 images with a size of $181 \times 217 \times 181$ voxels. All of these images are provided with ground truth segmentations for quantitative evaluations.

We exploit the structured-sparse RIT here to make predictions on the super-voxel level. To note, RIT is implemented on the semi-supervised version and, on each image, a random set of 150 super-voxels on each MRI are labeled with their ground truth label. In medical image segmentations, rather than the prediction accuracy, the dice similarity coefficients (DSC) is instead used as a criteria to ssay a method's performance [45][2].

For comparison purpose, the SS-RIT is pit against other leading methods for MRI segmentation including brain tissue segmentation algorithms in FMRIB Software Library (FSL) [46] and Statistical Parametric Mapping 8 (SPM8) package [47]. These two tools perform voxel-vise segmentation and have been widely regarded as benchmark methods in the neuroimaging community. Further, on the super-voxel level, the SS-RIT model is compared with other subspace models including MIM and TSC. MIM is also an information theoretic embedding method and TSC achieves much robust results according to previous tests. Finally, the structured sparse model is compared with RIT and Information Theoretic Discriminative Segmentation (ITDS) [38]. The major differences between ITDS and SS-RIT is the former exploits the ℓ_1 sparse norm for feature selection in the original space while the later conducts group-level feature learning in subspaces.

The segmentation results are visualized in Fig 3.6 Each panel illustrates the brain tissue segmentation result of the same brain volume selected from the IBSR dataset. The color of red, green and blue voxels represent the tissues of CSF, GM and WM, respectively. By comparing the segmentations to the ground truth, the advantage of SS-RIT segmentation over other six approaches is apparent. In particular, the SS-RIT segmentation shows particularly better delineations of CSF and GM tissues.

The quantitative evaluations on two datasets are reported in Table 3.5. A higher value of DSC represents a better correspondence to the ground truth. From the results, it is noted that the super-voxel level segmentations (last five rows in Table 3.5.) are much better than the voxel level segmentations (FSL and SPM8). Among all the subspace segmentation methods, the performance of RIT is much robust than MIM and TSC. This is because RIT conquers the uncertainties in the supervised labels and thus better captures the coherent discriminative structures in MRI data.

Further, we will discuss the advantages of exploiting structured-sparse norm which can be verified by comparing SS-RIT with RIT (no feature selection) and ITDS (sparse selection). Within the same experimental setting, learning group-level feature transformations (structured subspace learning) generally outperforms other two

---

[2] DSC is calculated from true positive (TP), false positive (FP) and false negative (FN) rates as, $DSC = \frac{2 \times TP}{2 \times TP + FP + FN}$.

Table 3.5 Performance of different segmentation methods on IBSR and BrainWeb datasets

Methods	IBSR			BrainWeb		
	CSF	GM	WM	CSF	GM	WM
FSL	0.53	0.76	0.87	0.85	0.88	0.90
SPM8	0.55	0.80	0.86	0.86	0.89	0.91
MIM	0.52	0.79	0.80	0.83	0.85	0.86
TSC	0.58	0.80	0.84	0.88	0.90	0.90
ITDS	0.60	0.81	0.86	0.90	0.90	0.92
RIT	0.63	0.83	0.87	0.92	0.92	0.93
SS-RIT	0.68	0.86	0.88	0.93	0.93	0.95

methods on DSC values. In addition, SS-RIT also achieves the lowest standard deviation for three types of tissues than others. This serves as an evidence to demonstrate that structured sparse learning does not only improve the accuracy but also enhances the robustness.

3.6 Conclusion

This work introduces an information theoretic method that successfully alleviates both label and feature uncertainties in general data representation tasks. The main advantages of the proposed method are represented from its flexibility and robustness. In the view of flexibility, RIT model works friendly with different types of feature transformation functions to conduct information theoretic learning. In this paper, we have implemented linear RIT, deep RIT and structured-sparse RIT models to address different image analysis tasks. The RIT framework generally improves other similar methods in the field. In the view of robustness, RIT is proven to be much effective to reduce the ambiguities in the training samples. Both the linear and deep RIT achieve much reliable performances in spite of label noises in the training samples. Moreover, its structured extension well addresses the partial volume effect (label uncertainty) and bias field effect (feature uncertainty) in MRI of brain tissue and thus achieves much better segmentation results than other state-of-the-arts.

In the model, the method focuses on mutual confirmation between the features and the predicted label, but it lacks the overall perception of the expected distribution of different types of samples. When there are large deviations in the distribution of different types of samples, the method performance will be affected. Therefore, in future research, fusion of the overall distribution of samples is also a problem worth exploring.

References

1. Böhning D (1992) Multinomial logistic regression algorithm. Ann Inst Stat Math 44(1):197–200
2. Giraldo LGS, Principe JC (2013) Rate-distortion auto-encoders. CoRR abs/1312.7381
3. Russakovsky O, Deng J, Su H, Krause J, Satheesh S, Ma S, Huang Z, Andrej K, Khosla A, Bernstein M et al (2014) Imagenet large scale visual recognition challenge. Int J Comput Vis 1–42
4. Belhumeur PN, Hespanha JP, Kriegman DJ (1997) Eigenfaces vs. fisherfaces: recognition using class specific linear projection. IEEE Trans Pattern Anal Mach Intell 19(7):711–720
5. Moore B (1981) Principal component analysis in linear systems: Controllability, observability and model reduction. IEEE Trans Autom Control 26(1):17–32
6. Hardoon DR, Szedmak S, Shawe-Taylor J (2004) Canonical correlation analysis: an overview with application to learning methods. Neural Comput 16(12):2639–2664
7. Zheng W, Zhou X, Zou C, Zhao L (2006) Facial expression recognition using kernel canonical correlation analysis (kcca). IEEE Trans Neural Networks 17(1):233–238
8. He X, Yan S, Hu Y, Niyogi P, Zhang HJ (2005) Face recognition using laplacianfaces. IEEE Trans Pattern Anal Mach Intell 27(3):328–340
9. Yan S, Xu D, Zhang B, Zhang HJ, Yang Q, Lin S (2007) Graph embedding and extensions: a general framework for dimensionality reduction. IEEE Trans Pattern Anal Mach Intell 29(1):40–51
10. Wang R, Shan S, Chen X, Chen J, Gao W (2011) Maximal linear embedding for dimensionality reduction. IEEE Trans Pattern Anal Mach Intell 33(9):1776–1792
11. Deng Y, Dai Q, Wang R, Zhang Z (2012) Commute time guided transformation for feature extraction. Comput Vis Image Underst 116(4):473–483
12. Tenenbaum JB, De Silva V, Langford JC (2000) A global geometric framework for nonlinear dimensionality reduction. Science 290(5500):2319–2323
13. Lee DD, Seung HS et al (1999) Learning the parts of objects by non-negative matrix factorization. Nature 401(6755):788–791
14. Lee H, Battle A, Raina R, Ng AY (2007) Efficient sparse coding algorithms. Adv Neural Inf Process Syst 19:801
15. Deng Y, Kong Y, Bao F, Dai Q (2015) Sparse coding-inspired optimal trading system for hft industry. IEEE Trans Industr Inf 11(2):467–475
16. Deng Y, Dai Q, Zhang Z (2011) Graph laplace for occluded face completion and recognition. IEEE Trans Image Process 20(8):2329–2338
17. Jenatton R, Obozinski G, Bach F (2010) Structured sparse principal component analysis
18. Mairal J, Bach F, Ponce J (2012) Task-driven dictionary learning. IEEE Trans Pattern Anal Mach Intell 34(4):791–804
19. Deng Y, Bao F, Kong Y, Ren Z, Dai Q (2016) Deep direct reinforcement learning for financial signal representation and trading. IEEE Trans Neural Networks Learn Syst 99:1–12. ISSN 2162-237X. https://doi.org/10.1109/TNNLS.2016.2522401
20. Deng Y, Ren Z, Kong Y, Bao F, Dai Q (2016) A hierarchical fused fuzzy deep neural network for data classification. IEEE Trans Fuzzy Syst (99):1–1. ISSN 1063-6706. https://doi.org/10.1109/TFUZZ.2016.2574915
21. Si S, Tao D, Geng B (2010) Bregman divergence-based regularization for transfer subspace learning. IEEE Trans Knowl Data Eng 22:929–942. ISSN 1041-4347. https://doi.org/10.1109/TKDE.2009.126
22. Gomes R, Krause A, Perona P (2010) Discriminative clustering by regularized information maximization. Adv Neural Inf Process Syst 23:775–783
23. Guyon I, Elisseeff A (2003) An introduction to variable and feature selection. J Mach Learn Res 3:1157–1182
24. Peng H, Long F, Ding C (2005) Feature selection based on mutual information criteria of max-dependency, max-relevance and min-redundancy. IEEE Trans Pattern Anal Mach Intell 27(8):1226–1238

25. Torkkola K (2003) Feature extraction by non parametric mutual information maximization. J Mach Learn Res 3:1415–1438
26. Grandvalet Y, Bengio Y (2005) Semi-supervised learning by entropy minimization. Adv Neural Inf Process Syst 17
27. Newton MA, Raftery AE (1994) Approximate bayesian inference with the weighted likelihood bootstrap. J Roy Stat Soc: Ser B (Methodol) 3–48
28. Donahue J, Jia Y, Vinyals O, Hoffman J, Zhang N, Tzeng E, Darrell T (2013) Decaf: a deep convolutional activation feature for generic visual recognition. *arXiv preprint* arXiv:1310.1531
29. Deng Y, Dai Q, Liu R, Zhang Z, Hu S (2013) Low-rank structure learning via nonconvex heuristic recovery. IEEE Trans Neural Networks Learn Syst 24(3):383–396. ISSN 2162-237X. https://doi.org/10.1109/TNNLS.2012.2235082
30. Jenatton R, Audibert JY, Bach F (2011) Structured variable selection with sparsity-inducing norms. J Mach Learn Res 12:2777–2824
31. Georghiades AS, Belhumeur PN, Kriegman DJ (2001) From few to many: Illumination cone models for face recognition under variable lighting and pose. IEEE Trans Pattern Anal Mach Intell 23(6):643–660
32. Nene SA, Nayar SK, Murase H (1996) Columbia object image library (coil-100). Technical Report CUCS-006-96
33. Lee KC, Ho J, Kriegman D (2005) Acquiring linear subspaces for face recognition under variable lighting. IEEE Trans Pattern Anal Mach Intell 27(5):684–698
34. van Gemert JC, Veenman CJ, Smeulders AWM, Geusebroek JM (2010) Visual word ambiguity. IEEE Trans Pattern Anal Mach Intell 32(7):1271–1283
35. Oliva A, Torralba A (2001) Modeling the shape of the scene: a holistic representation of the spatial envelope. Int J Comput Vis 42(3):145–175
36. He L, Parikh NA (2013) Automated detection of white matter signal abnormality using t2 relaxometry: application to brain segmentation on term mri in very preterm infants. Neuroimage 64:328–340
37. Greenspan H, Ruf A, Goldberger J (2006) Constrained gaussian mixture model framework for automatic segmentation of mr brain images. IEEE Trans Med Imaging 25(9):1233–1245
38. Kong Y, Deng Y, Dai Q (2015) Discriminative clustering and feature selection for brain mri segmentation. IEEE Signal Process Lett 22(5):573–577
39. Achanta R, Shaji A, Smith K, Lucchi A, Fua P, Su?sstrunk S (2012) Slic superpixels compared to state-of-the-art superpixel methods. IEEE Trans Pattern Anal Mach Intell 34(11):2274–2282
40. Liao S, Law MWK, Chung ACS (2009) Dominant local binary patterns for texture classification. IEEE Trans Image Process 18(5):1107–1118
41. Lowe David G (2004) Distinctive image features from scale-invariant keypoints. Int J Comput Vis 60(2):91–110
42. Felzenszwalb PF, Girshick RB, McAllester D, Ramanan D (2010) Object detection with discriminatively trained part-based models. IEEE Trans Pattern Anal Mach Intell 32(9):1627–1645
43. Rohlfing T (2012) Image similarity and tissue overlaps as surrogates for image registration accuracy: Widely used but unreliable. IEEE Trans Med Imaging 31(2):153–163
44. Kwan RK-S, Evans AC, Pike GB (1999) Mri simulation-based evaluation of image-processing and classification methods. IEEE Trans Med Imaging 18(11):1085–1097
45. Dogdas B, Shattuck DW, Leahy RM (2005) Segmentation of skull and scalp in 3-d human mri using mathematical morphology. Hum Brain Mapp 26(4):273–285
46. Zhang Y, Brady M, Smith S (2001) Segmentation of brain mr images through a hidden markov random field model and the expectation-maximization algorithm. IEEE Trans Med Imaging 20(1):45–57
47. Ashburner J, Friston KJ (2005) Unified segmentation. Neuroimage 26(3):839–851

Chapter 4
Computational Recovery of Sample Missings

Many problems in the real world include uneven label distributions where the proportion of samples of one type in the data is greatly different from the proportion of samples of other types. Due to the difficulty in the data collection, this data imbalance problem widely exists. For life science issues, this data sampling missing is difficult to alleviate by improving data-collection methods. Thus the problem of data imbalance poses a challenge to traditional machine learning methods. In response to this problem, this chapter proposes structure-aware and rebalancing learning. Through the analysis of internal structure of the data, the proposed method tries to rebalance the unbalanced data. On the association analysis and prediction tasks, we demonstrate the strucure-aware rebalancing method can efficiently improve the analysis of imbalanced data.

4.1 Introduction

In this chapter, we consider two tasks under the background of missing samples: data association analysis and data classification. Data association analysis is widely used in biological data analysis. The main target is to identify the causal association level between the features collected in the sample and the corresponding labels. Based on the results, we can identify important features that are strongly associated with label outcomes. A classical example is the genome-wide association study (GWAS), where both gene sequencing of each individual and disease status are known. Under this condition, the association analysis can identify gene loci that have significant contributions to the disease, thereby providing a basis for further functional analysis of the cause of the disease. Association analysis is an important statistical analysis method in biomedical research. Data classification is a more general problem. By analyzing the relationship between sample features and labels, the goal of optimally predicting data types can be achieved. Traditional association analysis and data clas-

© Tsinghua University Press 2021
F. Bao, *Computational Reconstruction of Missing Data in Biological Research*,
Springer Theses, https://doi.org/10.1007/978-981-16-3064-4_4

sification assume that the data is inherently balanced and there is no serious missing of a certain category. However, in practice, the lack of data is a widespread and unavoidable problem, resulting in sample imbalances in varying degrees in collected data.

On datasets with biased sample categories, the performance of existing statistical methods will have serious deviations, which poses severe challenges to the effective exploition of biological data and the reliability of biological conclusions. In the context of imbalanced learning, this chapter focuses on the data association analysis and data classification problems under the condition of missing samples.

For data association analysis, we take the genome-wide association study as the research application object. THE genome-wide association study (GWAS) plays an important role in revealing suspicious loci from statistical view [1]. However, the results of GWAS are still suffering from the missing heritability phenomenon [2], lacking sufficient powers to explain the causality of complex diseases. While a number of pioneering works have been proposed from different aspects to tackle the problem, their findings still hardly offer comprehensive heritability explanations either in the statistics, or the biology [3–5]. A number of inherent factors may decrease discovery powers of conventional GWAS methods on heterogeneous genomic data.

In this work, we focused on an important but always neglected risky factor for missing heritability: the imbalance of case/control samples. In GWAS, the acquisition of case samples is more difficult than obtaining the control data because the former can only be genotyped from individuals with the required phenotype, *e.g.* a certain cancer type. When performing GWAS on a rare disease [6] or an ethnic group, it is extremely hard or even impossible to collect enough case samples, yielding severe imbalance between case and control groups [7].

In GWAS, the logistic regression (LR) with χ^2 test (1 degree, $1 - df$) is typically used to identify the association between each locus and the phenotype [8, 9]. Compared with other GWAS methods, LR could take the effect of covariates (*e.g.* age, gender, smoke status, *etc.*) into consideration [8], and is hence becoming a benchmark tool in the field. It can be further extended to linear mixed models (LMMs) [10, 11], to include flexible adjustation to cofounder information such as population structures. However, directly conducting LR on imbalanced samples is apparently not reasonable. The samples in the case group (minor class) play less important roles than the control samples because the former only occupies a small portion of total training samples. This imbalance phenomenon can encourage the regression solution biased towards the major class [12]. As a result, the rich information about suspicious SNPs (hidden in the minor case group) is undervalued and consequently decreases the discovery ability of the LR method.

Herein, we introduce a novel approach, Bosco, to tackle the imbalance problem in association analysis. Bosco is constructed upon the existing boosting learning theory to assign samples in the major group with different important scores. However, the weighting strategy of Bosco is particularly designed to address the pursuit of suspicious SNPs discovery, while not on the straightforward classification side. Bosco is implemented in a coarse-to-fine learning framework. In the coarse step, we conduct the genomic level feature selection, *e.g.* ℓ_1-Lasso [13], on multiple randomly sampled and balanced subgroups. We then assess the quality of each subgroup by evaluating its SNP discovery consistency with other groups to assign a confidence score. In the fine step, the importance score for each sample is integrated to a weighted logistic regression [14], for SNP-level P-value calculation. Bosco was first evaluated through experiments on simulated data from the 1,000 Genomes Project [15]. Our approach showed better powers than the traditional scheme (classical $1 - df$ test and other imbalanced methods) and well controlled the false discoveries. The effects of Bosco were further highlighted on its discovery power with severe imbalance groups. We then applied the proposed method to analyze the gastric cancer dataset (with 745 cases and 2,100 controls). Our approach could replicate all the suspicious loci discovered by the conventional LR method with more significant P-values.

Learning from imbalanced data is a time honored problem in the machine learning community [16, 17]. Traditional approaches for imbalanced learning can be briefly divided into three categories including majority class under-sampling [18], minority class over-sampling [19] and cost-sensitive weighting [20]. However, these existing approaches generally fall into the shallow learning category that limits their potential adaptations in the current deep learning era. Compared with shallow learning, deep learning shows the ability to learn hierarchical feature representations by mimicking the supervised labels [21, 22]. However, to our knowledge, deep models for imbalanced learning are still not widely discussed. By analyzing the reason, we consider typical deep learning models being naturally not suitable for imbalanced learning because of their strong data fitting capability. The inputs to the neural network are biased imbalanced samples and the deep learning model is trained to optimally fit such imperfect labels. Conceivably, the trained discriminator does not own the robust generalization ability due to the "garbagein-and-garbage-out" problem.

Inspired by the shallow imbalanced learning approaches, we consider training a deep model with balanced training costs. Our main contribution is the proposition to learn and implement the rebalancing operations in a latent feature space. In detail, our approach first transforms the original imbalanced data into a latent space by a deep neural network. Then, it tries to discover two sets of equal-sized landmark points on both minority and majority classes. Finally, these equalsized landmark points are fed in a discriminative classier to seek for separable functions defined in the latent space. While we have interpreted these three major steps in a sequential manner, it worths emphasizing that these three respective steps are implemented in a joint deep learning framework. As a result, our learning framework allows feature learning, sample rebalancing and discriminative learning in an end-to-end trainable manner.

The major challenge in the aforementioned deep learning system is how to rebalance samples in the transformed latent space. We achieved this goal by conducting K-means operations on latent points in all classes. Then a set of K centers can be used as the landmark points that briefly represent the underlying data distributions in their respective classes. Due to the indispensable role of these latent landmark points, we termed the whole imbalanced learning framework as learning deep landmark in latent space (DELTA). It also worths noting here that while we mainly follow the undersampling concept to rebalance samples' distributions, alternative imbalanced learning approaches such as over-sampling and ensemble learning can also be incorporated into this basic DELTA framework.

DELTA does not only shed lights on the feature learning side but also makes a remarkable contribution in enabling imbalanced learning for data encoded with complex structures, such as trees and sequences. Existing imbalanced classification approaches mainly manipulate rebalancing implementations directly on the original data that hinders their applicability for structured data. This is because the similarity between two structured data is not easily measured when comparing them in the original data space. For instance, it is quite difficult to tell how to find K clustering centers for multiple parsing trees with diverse structures. It is meanwhile quite difficult to generate a new parsing tree by interpolating some observed trees in the minority group. However, these significant challenges are well handled by our DELTA model in its deep latent space. With appropriate feature extraction modules (such as a LSTM), the complex structured data could be easily encoded to latent vectors of the same length. Therefore, it is quite easy for DELTA to work on these numerical vectors with many rebalance operations in the DELTA latent space.

We show the effectiveness of DELTA model on many practical problems. First, we will compare DELTA with existing imbalanced learning models on several benchmark datasets that are extensively used by shallow learning works. Then, we consider adopting DELTA model in a more practical big data scenario for CTR prediction. Both a display and a mobile Ads datasets will be tested in this work. Finally, we discuss the indispensable ability of DELTA in handling structured data with imbalanced labels.

4.2 Related Works

Learning on imbalanced data is a common challenge faced in many practical problems ranging from image processing [17], to biological analyses [23]. Due to the imbalanced class distribution, methods developed on balanced data can suffer from severe loss of power when handling imbalanced problems. To enable the analysis on imbalanced data, a number of works have been proposed to tackle this imbalance learning problem [16, 24]. These works can be categorized to sampling-based methods and cost-sensitive methods.

Sampling-based methods focus on obtaining a balanced dataset from the imbalanced dataset using over-sampling or under-sampling [18] techniques. The random over-sampling and under-sampling are two basic methods in this category. By randomly replicating the minor samples or removing the major samples, the methods derive a balanced datasets from the original imbalanced data for the standard methods to process. Instead of simply replicating existing samples, Synthetic minority over-sampling technique (SMOTE) [19] creates synthetic minorities based on the existing sample and its neighbors, to take the local structure into consideration. For further modifications, the borderline-SMOTE emphasizes the samples near the class boundary and only uses them to simulate new minorities while the Adaptive Synthetic Sampling (ADASYN) estimates the data distribution first, and then creates new samples based on the distribution. Sampling with data cleaning methods, such as one-sided selection (OSS) and neighborhood cleaning rule (NCL), employ data cleaning techniques in the sampling process to remove the overlapping phenomenon. Cluster-based over-sampling (CBO) [25] takes each class as a cluster and defines the cluster border through an iteratively sampling process. Within the cluster border, CBO generates new minor samples to achieve a balanced dataset.

Cost-sensitive methods target on correcting the imbalance problem by placing different weights on samples, instead of sampling a balanced dataset [20]. One intuitive idea in this category is to apply different costs for different misclassification conditions. When minor samples are misclassified to major classes, the costs are much higher than the costs in contrary cases. AdaCost [26] modifies the weight updating methods in Adaboost to increase the cost for misclassification of minor samples. In decision three methods, cost-sensitive modifications have been applied in decision threshold [27, 28], split criteria and pruning. In neural networks, imbalance corrections has also been used in the learning process (modify the learning rate) and aim function (modify the output of neural networks and learning cost) [29]. Some algorithm-based methods also combine the sampling methods with the ensemble learning to solve the imbalance problem. EasyEnsemble builds balanced sub-datasets by bootstrap sampling and trains multiple classifiers on the sub-datasets to derive an overall result. Similarly, BalanceCasade [30] also employs multiple classifiers on ensembles but only uses the classifiers to select which major samples to undersample.

Conventional LR-based test in GWAS suffers from great bias when dealing with dataset with severe imbalance level. To enable LR to handle the biased dataset, many works were proposed to alleviate the imbalanced phenomenon. One intuitive idea is to generate synthetic samples of minor class based on existing samples. The synthetic minority oversampling technique (SMOTE) and its modified versions are the representative works in this category. However, most boosting approaches cannot be directly applied to the GWAS problem. This is because the learning objective in existing boosting method mainly focuses on improving the classification accuracy. However, the modification methods on GWAS aim to improve the significance level of causal markers, i.e. obtaining a smaller P-value under the same dataset. Improving the classification accuracy dose not promisingly lead to an improvement on significance test. They are actually two independent problems.

For classification problems, combining deep learning to achieve good performance is also the direction of current academic development. Deep learning has shown good performance in many tasks. Data imbalance is also an inevitable problem in this field. Intuitively, most sampling-based and cost-sensitive methods can be directly incorporated into the deep learning framework. Deep over-sampling (DOS) applies over-sampling technology in the hidden space defined by the deep neural network, and retrains the deep model on the balanced data set. Cost-sentitive deep neural network (CoSen DNN) [31] combines deep feature learning and cost matrix to automatically infer the optimal misclassification cost weight. The Class Rectification Loss (CRL) regularization algorithm designed a hard mining (Hard mining)

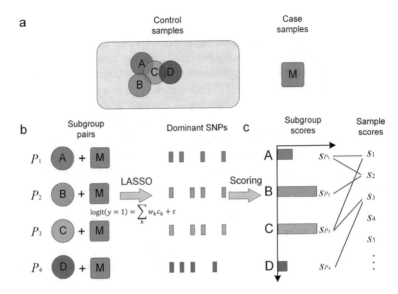

Fig. 4.1 The framework of Bosco. **a** Major samples are randomly divided as a number of overlapped subgroups (A, B, C, D). **b** Each major subgroup is combined with minor samples (M) to form a balanced pair. LASSO is applied to each pair for dominant SNPs selection with the additive genetic model. **c** A confidence score is assigned to each subgroup according to the appearing frequencies of dominant loci. Each samples confidence score is obtained by averaging all its appearing subgroups confidence score

scheme to emphasize the cost of each secondary category sample in each sample batch, and pay special attention to multiple label classification tasks. Large Margin Local Embedding (LMLE) [32] uses five-tuple sampling to achieve balanced input to the deep neural network, and limits three maximum hinge losses in each training cycle to constrain five yuan Group edge.

4.3 Structure-aware Rebalance Learning

4.3.1 Boosting Corrections for Genome-wide Association Studies with Imbalanced Samples (Bosco)

An overview of Bosco is illustrated in Fig. 4.1. Firstly, we divided the major group as L overlapping subgroups with the same size as the minor class, *i.e.* case group (labeled as M in figure). The word overlapping implies that one sample is allowed to be divided into multiple sub-groups. Then, we have accumulated L pairs of balanced minor group and subgroup combinations ($P_1, \ldots, P_4$). Based each pair, we resorted to the additive genetic model to conduct the genomic level SNP selection

(Fig 4.1b). The sparse optimization method LASSO [13] was used to select a set of phenotype-dependent SNPs, which were named as dominant SNPs for simplicity.

It was conceivable that L sets of dominant SNPs accumulated from these L pairs might not agree with each other. We then assigned the confidence score to each subgroup according to its consistency with other groups (Fig 4.1c). The rationality of our scoring mechanism lies on the assumption explained as below. Due to the uniformly random selection strategy, we expect each subgroup may fully capture the whole population structure of the large control set with little outliers. Therefore, the dominant SNP discoveries from each subgroup should exhibit moderately small fluctuations. Based on this assumption, if one sub-group leads to very unique discoveries but not reported in other subgroups, such discoveries should be treated as false positives. These false positives are assumed to be caused by noisy samples and outliers in that subgroup. Therefore, low confidence scores were assigned to samples in the corresponding subgroup. As an individual sample was allowed to be simultaneously selected into multiple subgroups, multiple confidence scores could be assigned to the same sample. We averaged these scores to calculate the final samplelevel score for each individual.

We noted that the aforementioned dominant SNPs selection step was performed on all SNPs simultaneously and thus it was termed as coarse selection. The major goal of this coarse selection step is to assign a confidence score to each sample in the control group. With this confidence score, we could further conduct our fine selection step at the individual SNP level.

According to the aforementioned coarse selections and scoring implementations, each of samples in the major class has been assigned with a confidence score s_i ($i = 1, \ldots, K$). Then, the corresponding confidence score was normalized as a weight to be taken into a weighted logistic regression (WLR). The WLR was performed on the single SNP level and significance of each marker was calculated under the conventional likelihood ratio test.

Selecting dominant SNPs from the genetic additive model
Subgroups were formed by randomly selecting samples from the major group with replacement. We repeated the sampling strategy a number of times to ensure the subgroups could cover most major samples. Each assembled subgroup is combined with minor samples to form a balanced combination (Fig 4.1b). On each pair of the combination, we construct a genetic additive model to characterize the contribution of each SNP.

$$\log (y = 1) = \sum_i w_i c_i + \varepsilon \qquad (4.1)$$

where the c_i is the ith SNP in genotype and w_i is its corresponding effect contributed to the phenotype y.

In the genetic additive model, SNPs were coded in 0,1,2, which indicated the number of minor allele(s) in loci. Each SNP (c_i) contributed to the disease linearly with the effect size defined by w_i. In this equation, the effects of multiple SNPs were considered simultaneously and the effect of each marker was determined together by solving this equation. Therefore, we could select those SNPs with large effects from the regression result. In the implementation, the regression includes millions of markers and only thousands of individuals, leading to an under-determined equation. We follow the same idea [33] to use LASSO (Least Absolute Shrinkage and Selection Operator) with ℓ_1-norm penalty to get a sparse estimation. After the LASSO estimation, we further select out T SNPs with the largest absolute effect weights w_i to form the dominant SNP set.

Scoring for subgroups and samples

After selecting out dominant SNP sets from L pairs, we evaluate the score of tth dominant SNP from the lth pair (see. Fig 4.1), *i.e.* $c_t^{(l)}$, by its frequency .

$$f_{c_t^{(l)}} = \sum_{i=1}^{L} I(c_t^{(l)} \in P_i)/L \tag{4.2}$$

where $I(\cdot)$ is the indicator function. P_i defines the ith pair. The frequency $f_{c_t^{(l)}}$ explains how often $c_t^{(l)}$ was identified from different pairs. A larger $f_{c_t^{(l)}}$ implies a higher confidence for the discovery of a suspicious SNP $c_t^{(l)}$ because it is identified from multiple sub-datasets. Then, we average $s_{P_i} = \sum_{t=1}^{T} f_{c_t^{(l)}}/T$ to assign a confidence score to the subgroup l, where T is the total number of dominant SNP in the set. As we assemble the subgroups in an overlapping manner, a same sample can be partitioned to many different pairs. Therefore, the confidence score for each sample is calculated as the mean score of all pairs where the sample is used.

A weighted logistic regression solution for imbalanced data

We consider there are K samples in the major group and k samples in the minor group ($K > k$). In addition, according to the aforementioned coarse selection and scoring implementation, each sample in the major class has been assigned with a confidence score s_i ($i = 1, \ldots, K$). Then, we transform the corresponding confidence score as a weight that will be used in the subsequent weighted logistic regression. For the ith sample, its weight is defined as $w_i^{(control)} = s_i / \sum_{j=1}^{K} s_j (i = 1, \ldots, K)$. The weight for the samples in the minor group (case) is just set to $w^{(case)} = 1/k$. To note, with these normalization strategy, the samples total contribution in these two respective groups have been normalized to one. We further employ the weighted logistic regression to model the association effects. The weighted logistic regression model resembles of the classical logistic regression but adds the weight for each individual sample.

$$L_w(\theta) = - \sum_{i=1}^{K+k} w_i \ln\left(1 + e^{(1-2y_i)X_i^T \theta}\right) \tag{4.3}$$

where y_i is the disease status of sample i, X_i is the factors in the regression including covariates and the tested SNP and w_i is its weight. θ is the parameter vector to be estimated. The parameters in $L_w(\theta)$ can be solved in the same manner as $L(\theta)$. After

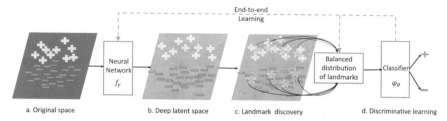

Fig. 4.2 An overview of learning deep landmark in latent space (DELTA). The feed-forward step of DELTA learning is indicated by arrows from left to right. **a** The original data space with imbalanced data (minority positive and majority negative). **b** Data distribution in a transformed latent space by a neural network. **c** Landmark points discovery by clustering to make a balanced number of landmarks on both classes. **d** Training a discriminative classifier with the balanced landmark points in the latent space. The optimization follows an inverse path (from right to left) to back-propagate the gradients from the discriminative loss to update parameters in both the transformation neural network and the discriminative classier

parameter estimation, we further employed the likelihood ratio test to conduct the significance test [34]. This test compared the statistical difference between $L_w(\theta)$ and the likelihood of logistic regression calculated without considering tested SNP. Then the significance statistic was defined as:

$$LR = \log L_w(\theta) - \log L_w(\theta'|NULL) \tag{4.4}$$

where $\log L_w(\theta'|NULL)$ denoted the likelihood computed only with covariates. According to the Wilks' theory, $2LR$ follows the χ^2 distribution with 1 degree-of-freedom. Thus the P-value was obtained by comparing the statistic with the distribution.

4.3.2 Learning Deep Landmark in Latent Space (DELTA)

Our approach for learning deep landmark in latent space (DELTA) was shown in Fig. 4.2. Without the loss of generality, we introduce our DELTA model in the binary situation. After understanding its binary setting, it is just straightforward to extend it for handling multiple classes. The algorithm of using DELTA for multiple classes will be discussed at the end of this subsection. We consider N^+ positive samples $x_i^+ \in \Omega^+$ and N^- negative samples $x_i^- \in \Omega^-$, $N^+ < N^-$ for imbalanced classes. Ω^+ (resp. Ω^-) represents the space spanned by positive samples (resp. negative samples). In the DELTA model, we believe there exists a mapping function $f_\gamma(\cdot)$ parameterized by γ that transforms the original data ($\mathbb{R}^p$) into a latent space ($\mathbb{R}^q$).

$$z_i = f_\gamma(x_i), \forall x_i \in \Omega^+ \cup \Omega^- \tag{4.5}$$

We noted that this generic transformation is applied to both positive and negative samples. In other words, such transformation is shared by all data points in all classes. There are many options to implement such parameterized transformation including linear mapping, kernel method and neural network [35]. In this paper, we choose neural network to encourage feature hierarchy in the data transformation process. In Fig. 4.2, we use the purple (resp. blue) color to represent the original (resp. latent) space before (resp. after) deep mapping.

After deep transformation, the imbalance issue still exists in the latent space as observed in Fig. 4.2b. Therefore, we consider using clustering approach to get K representative points from both minority and majority classes. Such K centers are landmark points defined in the deep latent space. More importantly, they are of the same quantity across all classes (Fig. 4.2c). While there are some off-the-shelf clustering methods, they are not suitable to be directly used in our DELTA model. The difficulty in our approach stems from the requirement of clustering 'latent' points in the 'latent' space. It is observed from Fig. 4.2c, the latent clustering part bridges the former feature learning and the latter discriminative learning parts. Therefore, we should expect an easy parametric implementation of such clustering so that the gradient can be easily passed through in the back-propagation learning phase. Therefore, we choose K-means as our basic clustering function because of its elegant matrix-factorization-type objective [36]. Accordingly, we get.

$$\min_{M^c, s_i^c} \sum_{c \in \{+,-\}} \sum_{i=1}^{N^c} ||f_\gamma(x_i^c) - M^c s_i^c||_2 \tag{4.6}$$
$$s.t.\ s_i^c \in \mathcal{B},\ \mathbf{1}^T s_i^c = 1$$

where $z_i^c = f_\gamma(x_i^c) \in \mathbb{R}^q$; $\mathcal{B}$ defines the binary space that means any vector $v \in \mathcal{B}$ could only have binary entries in $\{0, 1\}$; $\mathbf{1}$ is a vector with all one entries. The optimized variables in Eq.4.6 include a matrix $M^c = [m_1^c, m_2^c...m_K^c] \in \mathbb{R}^{q \times K}$ composed of K k-means centers and assignment vector s_i^c that crisply assigns the data $z_i^c = f_\gamma(x_i^c)$ to one of the clusters. The superscript c defines the class association implying we should implement such $K-$means clustering in each class separately. In each class, the clustering centers in M^c are so-called latent landmarks. We indicate the differences between this clustering operation and the previous feature transformation operation. The deep transformation in 4.5 is generic for samples in all classes. However, such landmark points discovery process is class-specific that needs to be performed on each class separately.

The promises of using clustering methods to balance minority and majority classes have been widely witnessed in a number of previous works [25]. However, we should emphasize that existing approaches just implement clustering on original data, while DELTA conducts clustering on latent points in a deep space. There are two apparent benefits of performing clustering in the latent space. First, deep transformation could effectively alleviate noises contaminate on the original data. Besides, deep transformation could encourage a $K-$means friendly data organization in the latent space. In detail, the clusters' structures may not be easily identified in the original data space due to noises and structural variations. But we can still expect their compact $K-$means organizations in the transformed space by designing a collaborative learning objective connecting the deep transformation and clustering functions [37].

After obtaining K latent landmarks in both positive and negative classes, it is easy to directly use such balanced landmarks as surrogates to fit a discriminative classifier as in Fig. 4.2d.

$$\min_{\theta} \sum_{c \in \{+,-\}} \varphi_\theta(M^c, Y^c) \tag{4.7}$$

where θ defines the parameters of the discriminant classier φ that assigns positive (M^+) and negative landmarks (M^-) to their respective class Y^+ or Y^-. Such a discriminate classier could also be implemented by a neural network with differentiable loss. Here, we will use cross entropy loss as the default choice. Till now, we have sequentially introduced three critical steps in DELTA model: 1) deep transformation (Eq. 4.5) , 2) landmark discovery (Eq. 4.6, and 3) discriminative learning on balanced landmarks (Eq. 4.7). It worths noting that none of the aforementioned three steps could be trained separately due to the lack of certain loss for each individual step. The loss could only be defined after seeing the classification errors from the final discriminative classifier (the rightest panel in Fig. 4.2). Accordingly, we integrate all these three steps in a joint framework and train the DELTA in an end-to-end manner. We get our learning objective as:

$$\min_{M^c, \gamma, \theta} L = \sum_{c \in C} \varphi_\theta(M^c, Y^c) + \lambda \sum_{c \in C} \sum_{i}^{N^c} ||f_\gamma(x_i^c) - M^c s_i^c||_2 \tag{4.8}$$
$$s.t.\ s_i^c \in \mathcal{B}, \quad \mathbf{1}^T s_i^c = 1, \forall c \in C$$

Here, we described the learning objective in a more general multi-classes scenario rather than the binary case. In (4.8), C defines how many classes are involved in the classification task. If $C = \{+, -\}$, it exactly corresponds to the binary classification problem discussed in Fig. 4.2. More generally, it also extends to serve multi-classes imbalanced learning problems when defining $C = \{1, 2, ..c\}$.

Optimization

We integrate learning targets of $K-$means and classification in a joint framework by using matrix factorization $K-$means. Consequently, Eq. 4.6 can be directly optimized.

The optimization of DELTA is not trivial because the problem in 4.8 induces to a highly non-convex objective. To solve non-convex optimization, we need to first find good initializations for all parameters in the system. We consider DELTA is composed of $N_\gamma + N_\theta$ layers, where N_γ, N_θ are the layer numbers of $f_\gamma(\cdot)$ and $\varphi_\theta(\cdot)$. First of all, we connect $f_\gamma(\cdot)$ and $\varphi_\theta(\cdot)$ altogether as a joint deep neural network and train it on imbalanced data. Then, all parameters in $f_\gamma(\cdot)$ and $\varphi_\theta(\cdot)$ can be readily used to initialize the corresponding part in the DELTA system. The last hidden layer in $f_\gamma(\cdot)$ is regarded as initialized latent sample in the deep transformed space. $K-$means algorithm can be easily performed on these latent samples to find all initial landmarks M^c in each respective class c.

With all parameters well initialized, we fine-tune the whole DELTA learning system in an end-to-end manner that was indicated by the dotted arrow in Fig. 4.2. We group learned variables into two sets: 1) neural network parameter set $\mathcal{X} = \{\gamma, \theta\}$ that include parameters in the mapping network f_γ and the discriminant classier φ_θ and 2) clustering related parameter set that include landmark matrix M^c and each individual assignment vector s_i^c, $\forall c \in C$. We follow the alternative way to update parameters in these two individual groups iteratively. We follow the alternative way to update parameters in these two individual groups iteratively. Here, we take M^c and s_i^c as parameters, because they will be updated continuously as the hidden space representation is updated.

When clustering parameters are fixed, the subproblem with respect to network parameters $\mathcal{X} = \{\gamma, \theta\}$ can be easily updated by gradient descent. We update this part of parameters by:

$$\mathcal{X} \leftarrow \mathcal{X} - \alpha \partial_{\mathcal{X}=\{\gamma,\theta\}} L, \tag{4.9}$$

where L is the learning objective defined in Eq. 4.8. In fact, such gradient information in 4.9 is easily calculated by Tensorflow. We use the ADAM optimizer [38] in Tensorflow to update the neural network parameters.

When the neural network parameters are fixed, we update the parameters related to the clustering operation in the latent space. First, we update the assignment one hot vector s_i^c in each class c as:

$$s_{j,i}^c = 1 \iff .j = \arg \min_{k=1..K} ||f_\gamma(x_i^c) - m_k^c|| \tag{4.10}$$

where $s_{j,i}^c$ means the jth element in the assignment vector s_i^c. When all s_i^c have been updated, the updating of the kth center m_k^c in class c is can be easily implemented by the averaging operation weighted by the assignment vector. However, such averaging approach may too much rely on the initialization of the assignment vector s_i^c. Therefore, they suggest an online learning strategy to avoid the aforementioned numerical problems by iteratively updating m_k^c via

$$m_k^c \leftarrow m_k^c - (1/n_k^c)(m_k^c - f_\gamma(x_i^c))s_{k,i}, \tag{4.11}$$

where n_k^c is the number counting how many times the learning process has assigned a sample to center m_k^c before handling the sample x_i^c. The superscript c implies that the clustering related parameters are processed at the perclass level. We show the detailed algorithm of DELTA learning in Algorithm1. In the equation, $(m_k^c - f_\gamma(x_i^c))s_{k,i}$ calculates the depth between the kth center m_k^c in the c category and the sample x_i^c with the label $s_{k,i}$ and then we average the value of n_k^c to update m_k^c.

Extensions to ensemble framework

From our previous discussions, the learning of the deep transformation function f_γ is inherently related to the classification loss φ_θ used in Eq. (4.8). In our previous statements, we chose the crossentropy loss as the discriminative objective. However,

Fig. 4.3 An overview of ensemble DELTA

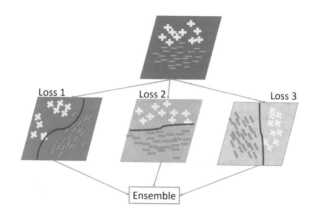

it is also very intuitive to understand that different supervised losses can guide the transformation function f_γ to find different γ as optimal parameters. As shown in Fig. 4.3, starting from the same data in the purple space, the learned representations in the latent spaces (those blue panels) are totally different when adopting different losses. Further, data points in different latent spaces are partitioned by different classifiers (red curves in different spaces).

We can then consider a simple approach to combine these different DELTA models altogether forming a strong classifier by borrowing the concept of ensemble learning. As shown in Fig. 4.3, in the training phase, we consider training each base DELTA model individually. Then, in the inference phase to classify a new data point, we respectively use these base classifiers to classify the data in their DELTA deep spaces. There are chances that each base classifier holds different opinion towards the classification result. We can simply find out the ensemble classification result by majority voting. In this paper, we consider three mostly used losses as our discriminative learning objectives including the cross-entropy loss, L_2 loss and the max-margin hinge loss.

4.4 Experiments

In this part, the performance of the proposed Bosco and DELTA will be verified through actual data. The different problems solved by the two methods are described separately.

4.4.1 Bosco's Assocaition Analysis on Imbalanced Data

Evaluation metrics on association analysis

In our work, we evaluated the performance of Bosco from two aspects: (i) the ability to reject false associations (type I error rate) and (ii) the ability to uncover true associations (empirical power).

Mathematically, the power and error rate indicate the probability that the significance test rejects the null hypothesis (no association between disease and SNP). In the first situation, we forced the effect size γ to 0, so that the disease status was generated with no relation to the genotypes. Any SNPs found associated under this assumption were false discoveries. In the second situation, we randomly selected one SNP as causal marker (c_i) and generated the phenotype according to Eq. 4.13 with an assigned effect size. In this manner, we got the ground-truth that SNP c_i was associated with y_i. Therefore, the ability of uncovering the true causal locus could be assessed.

In the calculation of the power (error rate) for SNP c_i, we randomly selected a sample group of n individuals G_k from the simulated dataset, where k indicated the index of the sample group. On the sample set G_k, we computed the significance P-value for c_i, denoting as $p_i^{(k)}$. We considered $k = 1, ..., K$ and selected different sample group G_k to compute the corresponding P-value for c_i on the new sample set. Then we define

$$U = \frac{1}{K} \sum_{k=1}^{K} I(p_i^{(k)} < \alpha) \tag{4.12}$$

where α was the significance level and $I()$ was the indicator function. If SNP c_i was the causal SNP (situation (i) aforementioned), then U was the empirical power; if not (situation (ii) aforementioned), U was the type I error rate. In the this work, we set for calculation as recommended in $K = 1,000$ in [39]. Typically, the α takes 0.05. However, in the situations where a number of hypotheses need to be tested, α should be corrected to $0.05/m$ for m hypotheses (Bonferroni correction).

In the simulation, we took 1,000 different SNPs to conduct the empirical power study independently. A power was obtained for each of the causal markers. Therefore, we present the average results among these 1,000 loci.

Preparation of simulation dataset

To evaluate the performance of our method in error rates and powers, we generated simulated genotypes based on clean data (without certain phenotype) and modelled the disease by designating causal SNP. For the simulation dataset, we randomly selected a region of 50M bp from the chromosome 1 as reference, based on the haplotypes of Utah Residents with Northern and Western European ancestry cohort (CEU) populations in 1,000 Genomes Project [15]. This region spans from 103,274,727 to 159,022,837 and includes 100,000 SNPs in total. We put this reference genotype into HAPGEN 2 software [40] and simulated genotypes of 10,000 samples. Then we chose the causal marker and simulated the phenotypes through a linear regression model:

$$logit P(y_i = 1|c_i) = \alpha_0 + \beta X_i + \gamma c_i + \varepsilon \tag{4.13}$$

where y_i is the disease status of sample i with 1 for case and 0 for control. X_i are covariates such as genders, weights, ages etc. c_i is the causal SNP that we designate before simulations. It contributes to the disease with effects controlled by the parameter γ (*a.k.a* effect size). We had two hypotheses: (i) a null hypothesis that no SNP had associations with the phenotype; (ii) an alternative hypothesis assuming the causal SNP contributed to the risk of the disease with an effect size of 0.5 ($\alpha_0 = 0.1$, $\beta = 0$, $\gamma = 0.5$). Then we simulated the case-control status using a regression framework under null or alternative hypotheses.

Type I error rate and empirical power studies on simulated data

In the simulation, we set the imbalance level to 0.2, which was defined as the proportion of the sample numbers between cases and controls, and randomly selected samples form the simulated population given the major sample size K (500, 1,000, 3,000 and 5000).

For Bosco, we set the size of major subgroups to be the same as the minor sample size k (that was 100, 200, 600 and 1,000 corresponding to each K). The number of subgroups was determined by $K/k \times 10$ to attain a trade-off between the sample coverage and the computation cost. For each subgroup pair, we used LASSO [13] to select the 100 top SNPs as dominant loci for scoring. We evaluated the method from two aspects, type I error rates and empirical powers, using the data generated under the two aforementioned hypotheses.

The results corrected by Bosco were compared with the results obtained under the classical regression method with likelihood ratio test (we used the Plink software [41] to conduct the regression test), oversampling strategies SMOTE [19], borderline-SMOTE (b-SMOTE), ACID [34] and BEAM3 [42], as well as a naive weighting cost strategy (nwCost) [14].

For Plink, we used the officially provided software and ran under the logistic regression model (under the `-logistic` command). For SMOTE, we considered the 5 nearest minor samples to generate new minors. Similarly, bSMOTE was configured by accounting 5 nearest neighbors regardless of their labels. For ACID and BEAM3, we followed the same setting in original publications. Finally for nwCost, the weights for samples in two classes were set as the $1/N$, where N denoted the sample number of the same class. Then the regression was solved under weighted logistic regression manner [14] and tested under the same likelihood ratio test.

We performed simulations on 1,000 random-selected SNPs independently. The type I error rate and empirical power were obtained under different significance levels ($\alpha = 5e - 02, 5e - 04, 5e - 06, 5e - 08, 5e - 10, 5e - 12$). For the type I error rates, Plink exhibited an ideal performance in controlling the false positive errors (Fig. 4.4 left panel). As comparisons, SMOTE, bSMOTE, nwCost suffered from large association errors. Error rates after correction by ACID, BEAM3 and Bosco were slightly larger than Plink, yet still in a relatively low level. The sample size had little effect on the errors.

As for empirical powers (Fig. 4.4 right panel), results of Bosco were better than those of competitors especially Plink and BEAM3 under all conditions, indicating Bosco improved the performance of the traditional methods effectively. When the sample number was small (major samples = 500, 1,000), the results of Plink suffered severe loss of power. Under this sample size, imbalance correction methods all tended to enhance the association ability. Bosco maintained the best performance

Type I error rate and empirical power studies on simulated data

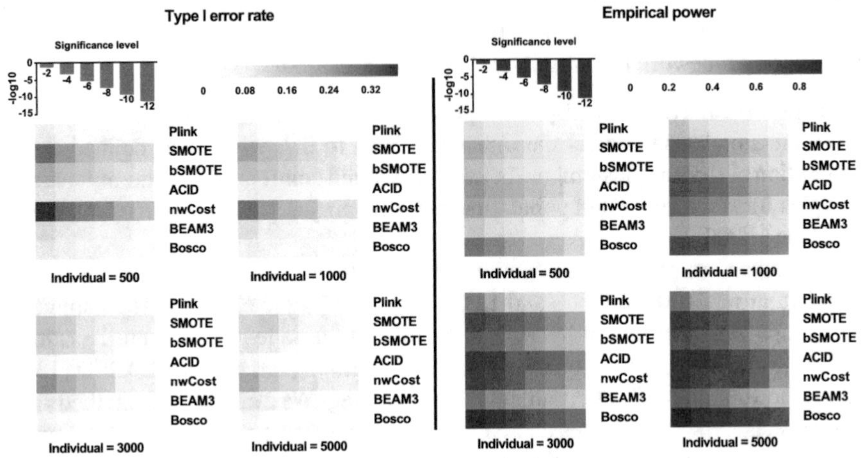

Fig. 4.4 The type I error rate (left) and empirical power (right) calculated under various significance levels and sample sizes. Under each experimental setting, the nominal type I error rate and empirical power were calculated under 1,000 repeats on different populations. In each color map, the significance level (α) was set as $5e - 02, 5e - 04, 5e - 06, 5e - 08, 5e - 10, 5e - 12$ from left to right

with significant advantages over Plink, bSMOTE and nwCost. When more samples were involved in the study, the performances of Plink got improved. Under this condition, the results of Bosco were still consistently better than others, especially under the significance level in typical studies involving millions of loci ($\alpha = 5e - 08$).

Combining the performances on both error rate and empirical power, we found most imbalance methods (SMOTE, bSMOTE and nwCost) failed to control the false positives, though they improved the powers with various levels. As a contrast, Bosco was effective in detecting associations and controlling the error rates.

The empirical power analysis on different imbalance levels

We further investigated the performances of Bosco in datasets with different imbalance levels. Experiments were also performed on 1,000 randomly selected suspicious SNPs with the effect size of 0.5. We varied the case/control proportion in the set of $[0.05, 0.1, 0.2, 0.3]$ The empirical powers were obtained under a variety of individuals (major sample = 500, 1,000, 3,000) with $\alpha = 5e - 04$.

When the imbalance level was small, *i.e.* 0.05 and 0.1, Bosco outperformed Plink as well as other imbalance methods greatly for all sample sizes (Fig. 4.5). Such result suggested Bosco could get reliable performances in severe imbalanced datasets. When the imbalance level was increased, the performances of Bosco and competitors got approaching. The simulation results indicated our method was particularly suitable to handle the tasks involving severely biased datasets and small populations.

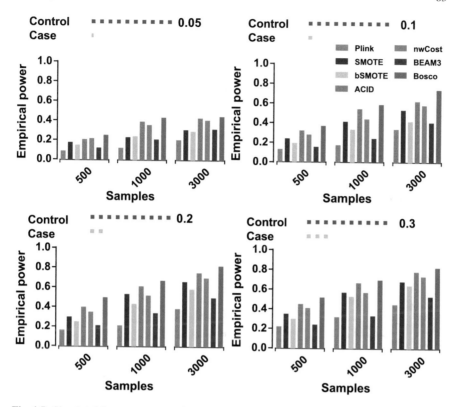

Fig. 4.5 Simulated data were generated under the effect size of 0.5. The control sample number was set to 500, 1,000 or 3,000 and case sample number was set according to imbalance levels. Powers were obtained under 1,000 experiments with the significance level of $5e - 04$

Simulation studies on the associations with different effect sizes

In association studies, different SNPs may contribute to the disease with different effect sizes. We further investigated the ability of our method in discovering suspicious SNPs with various association strengths. We followed the same protocol in the previous part and only changed the effect size of the suspicious SNP (in the set of $[0.1, 0.3, 0.5, 0.8, 1]$) to generate the simulated phenotypes. The empirical powers were obtained on 1,000 random-selected SNPs under the significance level of $5e - 04$. We reported the average scores of the empirical powers in Fig. 4.6.

In overall comparisons, Bosco achieved better average results on all effect sizes under different sample numbers (Fig. 4.6). When the sample size was small and the effect was weak, none of methods could accurately uncover the suspicious loci. However, when increasing the major sample number from 1,000 to 3,000, the empirical power could be improved dramatically, especially when the effect size was moderate.

Robust analysis on parameter settings for Bosco

In Bosco, different parameters may have potential influences on the performance of the method. We conducted analysis on two important parameters: the number of

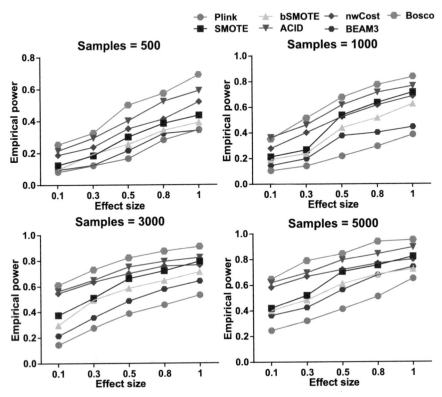

Fig. 4.6 The empirical power analysis on associations with a variety of effect sizes. The simulated phenotype was generated assuming associations were caused by SNPs with different effect sizes. Experiments were conducted on 1,000 different SNPs under various sample numbers. The imbalance level was set to 0.2

major subgroups (L) and the size of dominant SNPs (T). The simulations were also constructed in the same simulated data by setting the effect size, imbalance level and major sample number to 0.5, 0.2 and 3,000 respectively. The type I error rates and empirical powers were calculated under the significance level of $5e - 04$. In the experiments on subgroup numbers L, we followed the previous experimental implementation to set the subgroup size to 3000×0.2 and dominant SNP number to 100. Then we considered the subgroup number $L = m \times K/k$, where K and k were the control/case numbers. We altered m in the set of $[1, 2, 5, 10, 15, 20]$ and got the corresponding L in the set of $[5, 10, 50, 75, 100]$. From the results, we have observed that Bosco with small sample group number might attain a relatively larger type I error rate (Fig. 4.7a), while empirical power maintained high performance. When L was increased to a larger number, both indicators reached to relatively reliable states without further changing.

We further analyzed the performance of Bosco under different sizes of dominant SNPs (T). We set the subgroup size and subgroup number to 600 and 50 respectively.

Table 4.1 Performances of Bosco when different SNP numbers were considered in simulation

SNP numbers	1,000	10,000	100,000
Type I error rate	0.0212	0.0236	0.0209
Empirical power	0.6414	0.6528	0.6511

Table 4.2 Details about three studied datasets

	Class	Size	IR	Attributes
Vehicle	2	946	0.25	18
Patients	2	583	0.35	10
Diabetes	2	768	0.35	8

Then, we varied the dominant SNP number in the set of [10, 20, 30, 50, 80, 100, 200, 300]. Boscos performances on these data were summarized in Fig. 4.7b. For the type I error rate, results under different sizes of dominant SNPs had similar performances. However, the empirical power suffers fluctuation when the size of dominant SNP set was small (< 50). When more dominant SNPs were used, the power also approached to a reliable state.

We also conducted experiments assuming the covariates and bias had different levels of effects. We considered the bias (α_0) and covariate β had an effect size ranging from 0.1 to 0.5. Under each effect size, we computed the type I error rate and empirical power using the same simulation configurations as before. We found that different effect sizes had little effect on both the type I error rate and empirical power (Fig. 4.7c). As the significance test has already considered the bias and covariates, it had the ability to adjust to effects of different levels. Therefore, Bosco had stable performances under diseases partially caused by covariates.

Typical GWAS involves millions of SNPs. We extended the simulations to 10,000 and 100,000 SNPs to study the performance of Bosco when processing dataset with larger multiplicity. The experiments were conducted by setting effect size, imbalance level, major sample number and significance level to 0.5, 0.2, 1,000 and $5e - 04$, respectively (Table 4.1). For experiments on 100,000 SNPs, we only computed type I error rate and empirical power under 100 replicates due to the large number of significance tests. Bosco maintained stable performances across different SNP numbers.

4.4.2 The Performance of DELTA on Imbalanced Datasets

Evaluations on benchmarks

We consider evaluating the performance of DELTA on existing imbalanced datasets. Three widely used imbalanced datasets from the UCI machine learning

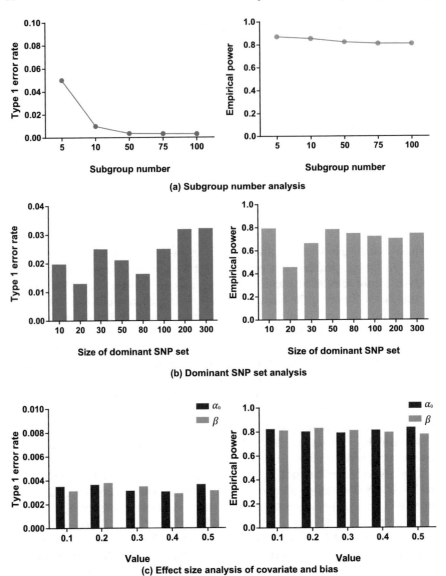

Fig. 4.7 The parameter analysis for Bosco. **a** the number of subgroups; **b** the size of dominant SNP set; **c** effect analysis on the covariate and bias. Simulations were constructed on 3,000 major samples with the imbalance level of 0.2 and the effect size of 0.5

repository will be tested including Vehicle [43], liver patients [44] and Prima Indian Diabetes datasets. The detailed information about these datasets can be found on the UCI website [45], and we just briefly summarized their properties in Table 4.2 including class number, sample size, imbalanced ratio (IR, defined by the percentage of the minority class samples to total data size) and their respective attributes number. While there are some other imbalanced datasets in UCI repository, they are not suitable in our DELTA scenario[1].

We compare our DELTA with other shallow and deep approaches. In shallow category, we choose random forest (RF) and XGboost as traditional supervised learning machines. These two ensemble learning algorithms are widely known to be robust to sample imbalances. They are directly trained on the imbalanced data without any balancing consideration. We further consider benchmark imbalanced classification techniques including Centroid [46] adaptive synthetic sampling (ADASYN), synthetic minority over-sampling technique (SMOTE) [19], Borderline SMOTE (BDSMOTE) and Majority Weighted MinorityOversampling Technique for ImbalancedData Set Learning (MWMOTE) [47], in which the first one is an under-sampling method and the later two are over-sampling approaches. After balancing training samples with these respective methods, a XGboost model is fitted on these balanced data. Those methods are tabulated in the "Shallow Learning" category in Table 4.3.

To our knowledge, there is no off-the-shelf deep-based imbalanced learning tool in the field. We hence consider some natural extensions of existing deep learning algorithms to make them applicable in handling imbalanced cases. The first method is to directly train a deep neural network with imbalanced data (DNN). Then, we consider using Borderline SMOTE (BDSMOTE) method to enrich the minority samples. Then, a DNN is trained with such balanced data (DNN+BDSMOTE). We use BDSMOTE as the sample balancing method because this certain approach is shown to be the best over-sampling method among all competitors in the "Shallow Learning" category of Table 4.3.

DNN is a supervised machine and its performance could be strongly affected by biased training labels. Therefore, we also consider auto-encoder (AE) as an unsupervised deep learning approaches to alleviate the heavy demands on labels. In detail, we train an auto-encoder to encourage the neural network to reconstruct the input data at the output side. The intermediate layer between the encoder and decoder is used as the latent feature. Then, we simply perform different oversampling approaches on those latent representations. After minority data augmentation, a XGboost classifier is trained to classify balanced samples in the AE latent space. In Table 4.3, we report performances by conducting both BDSMOTE and ADASYN to augment minority feature vectors learned from AE.

Next, we consider four off-the-shelf deep learning methods designed for imbalanced problems, including LMLE [32], CRL [40], CoSen DNN, and DOS [31]. For LMLE, we employed this method to learn deep representations from imbalanced data and then performed classification using $k$nearest-neighbor (kNN) as introduced

---

[1] Those datasets just contain less than 100 samples, which are not sufficient to be used for training a deep model.

Table 4.3 Imbalanced learning results on benchmark datasets(%)

Category	Methods	Vehicle	Patients	Pima
Shallow learning	RF	87.035±0.025	88.681±0.021	71.830±0.034
	XGBOOST	86.911±0.023	89.234±0.031	73.859±0.041
	Centroid	86.552±0.027	86.253±0.024	72.118±0.032
	SMOTE	86.624±0.024	88.216±0.031	76.236±0.042
	BDSMOTE	88.723±0.022	89.138±0.022	78.313±0.035
	ADASYN	86.755±0.028	88.054±0.024	77.325±0.041
	MWMOTE	86.428±0.023	87.382±0.032	76.388±0.048
	UCML	87.032±0.032	87.746±0.028	76.925±0.038
Deep learning	DNN	85.357±0.022	87.622±0.027	75.326±0.029
	DNN+BDSMOTE	86.532±0.028	89.339±0.031	76.886±0.032
	AE+BDSMOTE	87.324±0.031	87.853±0.027	78.353±0.027
	AE+ ADASYN	86.529±0.029	86.785±0.025	77.322±0.036
	LMLE	88.392±0.032	89.329±0.027	78.783±0.031
	CRL	88.182±0.023	88.627±0.029	77.327±0.029
	CoSen DNN	88.091±0.028	87.872±0.026	77.129±0.033
	DOS	87.926±0.027	87.231±0.023	77.922±0.024
DELTA learning	DELTA	88.320±0.020	89.867±0.029	78.884±0.034
	DELTA+BDSMOTE	89.892±0.024	90.063±0.030	80.523 ±0.035
	DELTA+ADASYN	88.732±0.021	89.780±0.022	78.321±0.038
	DELTA+XGBOOST	89.634±0.023	90.131 ±0.029	79.742±0.042

in the original research. For CRL, we used the instance level hard mining (CRL/I) as it reported the best overall performance in the original paper. These deep imbalanced methods are tabulated in the "Deep Learning" category in Table 4.3.

For DELTA related approaches, we first report the classification results by DELTA learning, which is essentially an under-sampling approach. Then we implement DELTA oversampling approach according to the method introduced before. Both the BDSMOTE and ANASYN are considered as over-sampling approaches conducted in the DELTA latent space. The classier in DELTA is trained with the binary crossentropy loss as the discriminative objective. We also consider training a XGboost machine with the balanced data points (obtained by BDSMOTE over-sampling) in the DELTA latent space. Results of these different approaches are reported in the "DELTA Learning" category of Table 4.3.

We chose parameters for comparing methods using the default values in packages or previous settings in orginal works. For XGboost, RF, SMOTE, and ADASYN, we directly employed default parameters in the python-based packages (for XGboost [2], maximal depth: 3, eta: 1, training round: 10; for RF [3], tree number: 10, minimum

---

[2] Refer to xgboost.readthedocs.io/en/latest/python/.

[3] Refer to scikit-learn.org.

sample number in leaf: 1, minimum sample in split: 2; for SMOTE and ADASYN,[4] number of nearest neighbors: 5). For all neural network related methods, we only consider one layer with eight units before the intermediate latent representation as the feature learning part. Also, only one layer with a sigmoid output is added after the latent representation layer for classification. The intermediate latent layer is fixed with eight units implying the hidden vector is in $\mathbb{R}^8$ latent space. We just consider such a slim network here because the original data dimension is quite small and the training sample size is also very limited as listed in Table 4.3. Such limitations on training data restrict the implementation of more complex network structures. For LMLE, we considered $k = 20$ nearest neighbors as suggested in the orginal work. In our DELTA approach, we set deep landmark size as $K = 50$.

We divide each dataset into ten equal-sized data folders. We then randomly select nine-folders of data to train respective learning models and the remaining one-folder of data was used for performance testing. Among nine-folders of training data, only eight-folders of data are directly used to train different models with the rest one reserved for hyper-parameter validation. We use area-under-curves (AUC) as a robust indicator to report the performances of imbalanced learning. AUC is defined in the range of (0.5, 1). For the binary case, AUC=0.5 means random guessing and AUC=1 implies perfect classification. The training and testing procedures are repeated for 10 times with the AUC evaluation results reported in Table 4.3 in the form of mean $\pm$ standard deviation.

From the results, we have observed that different methods perform quite similar on these three data sets although the best ones are all achieved by the DELTA-related approaches. The DNN and AE methods in the "Deep Learning" category do not improve the shallow methods with an obvious margin. However, we do observe a trend that all imbalanced learning approaches could improve the results of the same method trained directly from imbalanced data. The imbalanced deep learning methods robustly improved performances of DNN on imbalanced data sets. Among all three methods within DELTA category, we have observed that over-sampling-based DELTA learning is always better than the original undersampling-based DELTA algorithms. We employed K−means clustering as the default clustering initialization method in aforementioned results. As the learning objective of DELTA is highly nonconvex, well-initialized parameters will facilitate the model-training greatly. To provide better initializations, we also took two distribution-respecting clustering methods (hierarchical clustering and CLIQUE) for the initialization of cluster assignment (Table 4.3). Compared with results under random initialization, cluster-based initialization greatly improved the accuracies on all three data sets. DELTA initialized by two new clustering methods also showed better performances compared with K−means initialization, with higher accuracy and lower standard deviation. While we have observed advantages of DELTA on all these three simple data sets, the limitation of available training data, indeed, restricts the power of deep-learning-based approaches. In the following sections, we will consider more suitable big data scenarios with a relatively large amount of training data.

---

[4] Refer to imbalanced-learn.org.

Table 4.4 DELTA with different clustering initialization methods

Initialization	Vehicle	Patients	Diabetes
Random	86.214±0.037	88.326±0.032	76.786±0.041
K−means	88.320±0.020	89.867±0.029	78.884±0.034
Hierarchical clustering	89.132±0.017	89.882±0.021	79.329±0.027
CLIQUE	88.672±0.013	90.021±0.022	79.332±0.029

Table 4.5 Summary on two CTR datasets

Datasets	Sample size	Minority ratio	Attributes
Display	46 million	0.26	228
Mobile	40 million	0.17	100

Binary classification evaluation on large-scale datasets

Click-through-rate (CTR) prediction is the core of modern computational adver-
tisement (Ads) and recommender system. The predictor in the CTR system is required
to precisely identity whether a certain user will click on a recommended Ads or not.
Serving the right Ads to the right user will help the merchant accumulate huge rev-
enues. However, as experienced in our daily life, end-users naturally dislike too many
annoying Ads that were recommended to them. Therefore, when checking historic
records of any recommender system, it is easy to find out severe imbalances between
the number of users' clicks and non-clicks. When phrasing this CTR problem as a
machine learning task, it is not hard to conceive that there are a huge amount of
negative samples and limited positive samples in the training data.

In this work, we consider expand our experimental discussions on two practical
Ads datasets including a display-Ads dataset [48] and a mobile-Ads dataset [49].
Some attributes about the Ads in these two datasets are provided in the form of raw
categorical information, such as the "device type". These categorical feature are not
possibly directly used for consequent learning systems. We hence convert them as
binary vectors to indicate which category they belong to. The detailed feature (a.k.a
attributes) number after categorical conversion are listed in Table 4.5 as well as other
detailed information about these two Ads datasets. The machine learning task here
is to predict the clicking probability of an Ads item for a specific user. Both the Ads'
and user's information has been summarized in the feature vector. Therefore, we
can simply regard this CTR prediction task as a binary classification problem with
imbalanced labels.

We noted that Ads records in these two datasets are sorted in the ascending time
order that respectively cover 7 days and 10 days durations. The time-variant effects
in Ads recommendation should not be neglected because one hot Ads can quickly
lose its interests to the public in just a short period. Therefore, we strictly follow the
arrival orders of Ads and split them into training and testing sets. In other words,
it is not possible to train the system with latter records and predict the CTR in an

early time. To this end, we consider uniformly dividing each dataset into 20 groups in an ascending order. Each group contains 5% items accumulated in a consecutive period. The tests start from the sixth group of Ads items by using its previous five groups of data as training data. Then, the testing set is incrementally moved forward by using its five previous groups for training. Such progresses are iteratively repeated until finishing the CTR prediction on the 20th group, leading to a total number of 15 individual testing groups.

All neural networks used in our DELTA and other deep learning methods share the same configuration. In detail, they all have 2 layers as feature extractor with 128 and 64 units per layer. The intermediate representation layer has 64 units, followed which are a two-layers-classifier: one 64 units layer plus an extra layer with sigmoid output for binary classification. The number of landmark points K is fixed as 800. We compare different methods' performances on these two Ads datasets in Table 4.6. It is shown that deep-learningbased approaches generally perform much better than the shallow ones. It is mainly due to the data size have been significantly enriched so that we have enough "fuel" to boost these deep machines. Meanwhile, DELTA-based approaches further improve traditional deep learning by comparing the results between the "DELTA" and "Deep" categories.

In this CTR test, we also consider the extension of using Ensemble-DELTA model as introduced before. We simply implement this approach by training three respective DELTA models with cross-entropy, least square and maxmargin hinge losses. Then, results of these three DELTA models are voted with equal weight. We noted that among all three DELTA-based approaches, ensemble DELTA achieves the best AUC score on two datasets. All three basic models in this ensemble DELTA are trained with the under-sampling concept, i.e. K-means reduction in latent space. However, their joint ensemble performances can even beat the over-sampling DELTA approaches such as DELTA+BDSMOTE. This is a good evidence showing the great promises of using DELTA model to learn different feature representations via various discriminative losses.

A critical hyper-parameter in DELTA is the number of landmark points K. Therefore, we alter such parameter in a reasonable range to see how it affects the performance of DELTA model. We consider the basic DELTA model and its two extensions, i.e. over-sampling DELTA and ensemble DELTA. The results with different K numbers on two Ads datasets are reported in 4.8 From the comparison, we have noticed that the performances of different DELTA models are relatively reliable when K is larger than 800. When $K < 800$, all DELTA methods suffer from accuracy drop. Meanwhile, the ensemble DELTA is more reliable than others when the number K is small. Besides, two DELTA extensions both beat the basic DELTA model on two datasets with different K numbers. Ensemble DELTA is the best among the three.

In experiments above, we set a equal landmark point number for both major and minor classes. To illustrate how this uniform K setting alleviated the imbalance problem in learning, we further reanalyzed the results by using a smaller K for minor class. In detail, we set the number of minor landmark point to 600, 400, and 200, while fixing the K for major class to 800. From the results in Fig. 4.9, the accuracy of DELTA decreased a little when number of landmark point is slightly

Table 4.6 The AUC on two CTR datasets (%)

Category	Methods	Display	Mobile
Shallow	RF	66.21±0.21	63.56±0.32
	XGBOOST	67.06±0.34	65.41±0.30
	Centroid	65.38±0.26	64.39±0.29
	SMOTE	67.31±0.22	65.37±0.29
	BDSMOTE	69.12±0.28	68.05±0.25
	ADASYN	69.89±0.31	66.42±0.26
	MWMOTE	68.37±0.23	66.03±0.28
	UCML	67.47±0.28	65.86±0.27
Deep	DNN	68.82±0.27	67.41±0.28
	DNN+BDSMOTE	72.32±0.20	68.92±0.33
	AE+BDSMOTE	73.34±0.24	67.35±0.29
	AE+ADASYN	73.37±0.33	69.90±0.29
	LMLE	74.71±0.35	71.02±0.35
	CRL	75.23±0.28	69.87±0.33
	CoSen DNN	74.32±0.30	70.12±0.36
	DOS	73.55±0.26	69.23±0.27
DELTA	DELTA	75.85±0.34	70.21±0.30
	DELTA+BDSMOTE	76.60±0.28	72.33±0.29
	DELTA+ADASYN	76.02±0.26	71.54±0.31
	DELTA+XGBOOST	75.81±0.26	72.33±0.37
	DELTA Ensemble	77.31 ± 0.17	74.32±0.23

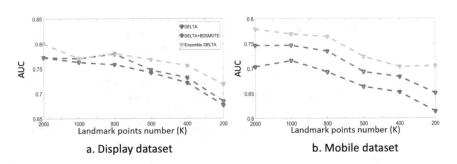

a. Display dataset   b. Mobile dataset

Fig. 4.8 DELTA learning with different number of landmark points (a.k.a K-means centers)

imbalanced (minor-$K = 600$). However, the accuracy could get dramatic drop when the imbalance level of landmark point numbers was large. When minor-K was set to 200, the proportion of major-K and minor-K was almost the same as proportion of major and minor samples. In this case, the performances of DELTA were still better than the results of shallow models.

Another case is users may emphasize the accuracy on certain classes in real applications. For example, the accuracy of minor samples (clicking ads) in CTR is

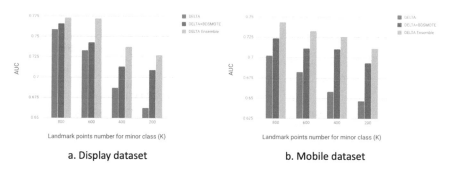

a. Display dataset   b. Mobile dataset

Fig. 4.9 DELTA learning with different number of landmark points for minor class. **a** Display data set. **b** Mobile data set

Table 4.7 AUC on CTR datasets wieh emphasis on positive samples (AUC %)

	Display		Mobile	
K	Major	Minor	Major	Minor
400	75.32±0.27	74.13±0.28	65.76±0.29	66.43±0.29
600	72.73±0.28	75.43±0.23	63.24±0.21	67.59±0.28
800	69.28±0.27	77.23±0.22	61.03±0.32	69.38±0.21

more important than predicting major samples. In this case, we can set a larger K for minor class to emphasize the accuracy for clicking ads. To illustrate this, we kept the K for negative samples to 400 and increased the K for positive samples from 400 to 600, 800. From Table 4.7, the accuracy for minor samples could get increased along with the increase in K. However, the accuracy for major samples got decreased.

4.5 Biological Application

4.5.1 Association Analysis of Genome Data on Gastric Cancer

To demonstrate the real application of Bosco, we conducted a genome-wide association analysis on the gastric cancer GWAS dataset from the database of Genotypes and Phenotypes [50] (dbGaP; phs000361.v1.p1). The gastric cancer dataset is part of the data from a meta-analysis [51]. Previously, this dataset belonged to an association study on gastric cancer (GC) and esophageal squamous cell carcinoma (ESCC) in ethnic Chinese subjects [52]. Samples were collected from Shanxi Upper Gastrointestinal Cancer Genetics Project (Shanxi) and Linxian Nutrition Intervention Trials (NIT) using Illumina 660W Quad chip. In our experiment, it included 2,100 control samples and 745 case samples, leading to a relative large imbalance

Table 4.8 The associated SNPs discovered by bosco in gastric cancer dataset of 2,100 controls and 745 cases. Plink discovered 5 suspicious loci and bosco replicated all of them with more significance, with 2 additional discoveries. Findings indicated a strong association between *Plce1* and the gastric cancer. ($\alpha = 2.23E - 07$)

Chr.	SNP	Gene	Pos.	Major/Minor	P Value		
					Plink	ACID	Bosco
1	rs4460629	Non-genetic region	153401959	C/T	2.17E-05	6.14E-07	3.45e-09
1	rs4072037	*MUC1*	153428691	A/G	3.41E-05	2.21E-07	2.09E-09
10	rs753724	*PLCE1*	96041407	G/T	8.01E-08	2.14E-10	6.37E-11
10	rs11187842	*PLCE1*	96042501	C/T	5.52E-07	3.66E-07	5.33E-10
10	rs3765524	*PLCE1*	96048288	C/T	3.59E-09	4.15e-09	7.32E-12
10	rs2274223	*PLCE1*	96056331	A/G	1.64E-09	8.98E-09	3.15e-09
10	rs3781264	*PLCE1*	96060365	T/C	6.97E-10	7.15e-11	7.36E-12

level (745/2, 100).After the standard quality control (QC) procedure (SNP call rate > 95%, minor allele frequency (MAF) > 0.05, and Hardy Weinberg Equilibrium (HWE) > 0.001), 223,823 SNPs were enrolled into the final association analysis.

Herein, we only employed Plink, ACID and Bosco for comparisons as other methods suffered severe error rates that the variants they identified had large possibility to be false positives. We set the parameters for Bosco using the same method (subgroup number = 2, 100/745 × 10 = 29; dominant SNPs = 100) and conducted the association test on each markers, with comparisons to the traditional logistic regression method (Plink) and ACID. On a computer with Intel i7 CPU and 16G RAM, Bosco took about 39 minutes to finish the genome-wide association analysis using multiple kernels in parallel, compared with 11 minutes of the latest Plink (ver. 1.9).

Bosco reported seven SNPs exhibiting the genome-wide significance under Bonferroni corrections (with P-value $< 0.05/223, 823 = 2.23e - 07$; Table 4.8). As comparisons, Plink only discovered five of these seven markers and ACID located six of them. For same variants detected by both methods, P-values computed by Bosco were much smaller than the results of Plink, showing the better power in discovering the same associations. The majority of suspicious loci were mapped to *PLCE1*, indicating a strong association between *PLCE1* and the gastric cancer.

Table 4.9 Cell type composition of retina dataset

Cell type	Cell number	IR*
Amacrine cells	4,426	0.102
Rods	29,400	1
Cones	1,868	0.043
Bipolar cells	6,285	0.144
Muller glia	1,624	0.037
Overall	43,603	–

*Here IR is defined as the class size ratio between the corresponding cell type to the major cell type Rods.

4.5.2 Cell Type Identification of Single-sell RNA-seq Retina Dataset

Learning meaningful cell features to help identify different cell types is an important task in biological research studies. However, the composition of cell types in biological tissues also exhibits imbalanced distributions. There are always several cell types dominating the cell population and the rest minor cell types only occupy a small proportion in the tissue. Here, we try to address this imbalance learning problem in cell composition. We employed gene expression levels as original features for each cell, as they are unique indicators of inner cell biological activities.

We apply DELTA to predict the imbalanced cell composition of one single-cell ribonucleic acid (RNA)-seq retina data set. There are 43, 403 cells in this data set and each cell has 384 genes to define cell attributes [53]. The data set includes five different cell types, with one major cell type (Rods) accounting for over 67.4% (Table 4.9). Comparing with the major cell type, two cell types show severe imbalance ratios smaller than 0.05 (cones and muller glia) while another two cell types show moderate imbalance levels between 0.10 and 0.15 (amacrine cells and bipolar cells).

On this data set, we followed the evaluation strategy as before and divided the data set into ten folders with equal size. Nine folders of data were used for training, while the left one was used for testing. This training-testing process was repeated ten times and AUC was used to evaluate the performance. For this task, we employed a three-layer feature extractor for deep learning and DELTA methods, with 128, 64, and 32 units in each layer. To achieve multiclass classification, cross entropy function was used to calculate the loss. Landmark number (K) were set to 500, for DELTA and its extensions considering the minor sample sizes. Ensemble-DELTA summarized the results of three DELTA models trained with cross-entropy, least square, and hinge loss.

Results of all comparing methods on the retina data set were illustrated in Table 4.10. For this data set, the deep-based models greatly outperformed the shallow models due to a deeper neural network structure was used for this task. Within the deep category, approaches specially designed for imbalance problems showed better performances than simple imbalance extensions based on convolutional neural

Table 4.10 AUC on retina dataset (AUC %)

Category	Methods	AUC
Shallow	RF	70.31±0.30
	XGBOOST	73.11±0.29
	Centroid	73.62±0.31
	SMOTE	75.21±0.25
	BDSMOTE	76.62±0.32
	ADASYN	75.42±0.35
	MWMOTE	74.32±0.31
	UCML	73.72±0.34
Deep	DNN	78.29±0.29
	DNN+BDSMOTE	80.52±0.34
	AE+BDSMOTE	83.93±0.31
	AE+ADASYN	83.37±0.35
	LMLE	85.02±0.34
	CRL	85.93±0.29
	CoSen DNN	84.12±0.31
	DOS	83.78±0.36
DELTA	DELTA	86.21±0.32
	DELTA+BDSMOTE	86.72±0.31
	DELTA+ADASYN	86.53±0.34
	DELTA+XGBOOST	86.13±0.28
	DELTA Ensemble	87.56±0.25

network (CNN) and AE. DELTA and its extensions further improved the accuracy and ensemble-DELTA, finally achieving the best performance with a 12% improvement compared with shallow models.

One important reason to explain the performance of DELTA framework is the latent DELTA space can separate different sample groups better after the deep transformation. To illustrate this, we visualized the sample distributions of original space and latent space in 2-D space by t-distributed stochastic neighbor embedding (tSNE) and evaluated the compactness of all classes using Calinski and Harabaz score [54] which was defined as

$$s = \frac{Tr(B_k)/(k-1)}{Tr(W_k)/(n-k)}. \tag{4.14}$$

In the equation, n is the sample size and, k is the class size. $Tr()$ is ttrace function to calculate the summation of elements in the main diagonal. B_k and W_k are sample convariance matrices between classes and within classes, which are defined as follows.

$$B_k = \sum_q n_q (c_q - c)(c_q - c)^T, \qquad (4.15)$$

$$W_k = \sum_q \sum_{x \in C_q} (x - c_q)(x - c_q)^T, \qquad (4.16)$$

where C_q is the set of all n_q samples in the class q and c_q is the class center (defined as the average of all class samples). c is the center of all samples. A larger Calinski and Harabaz score indicates samples within the same class have better compactness and samples between classes are more separable to each other. From the tSNE visualization (Fig. 4.10), minor classes have a larger margin to the major class in the DELTA latent space. The Calinski and Harabaz score on latent space is also significant larger than that in original space. This indicates that the compactness of latent space is greatly improved.

4.6 Conclusion

This chapter discusses a special aspect of missing data: overall missings of samples. In both natural image or in biomedical research, this problem commonly exists. From the perspective of structure-aware rebalance, this chapter provides solutions from two different data analyses: association analysis and classification prediction.

For the first type of analysis, we illustrated that the case/control sample proportion could also be a risky factor that potentially harmed the association power. One straightforward solution to this long-standing problem is to add more minor samples (either replicating exiting samples or synthesizing new samples) to encourage the data to reach a balanced state. However, using too many similar samples in the analysis will lead to the population stratification and cause spurious associations [55]. In this chapter, we proposed a new approach called Bosco to perform corrections for the imbalanced GWAS datasets. Bosco exploited a coarse-to-fine strategy to learn on a number of balanced pairs and then heuristically assigned an importance score to each sample in the control group. Both simulations and application demonstrated Bosco could greatly improve the powers in discovering associations from the biased dataset.

For the second type of analysis, classification prediction is a more general task. For the prediction on imbalanced data with missing samples, we introduce a DELTA model for imbalanced learning in the deep space. DELTA has three properties: latent transformation by deep neural network, reimbalancing by sample clustering, and training by a joint, end-to-end framwork, making it capable of achieving better performances in imbalance classfication tasks. More importantly, DELTA offers the possibility to conduct imbalanced learning on structured data due to its feature learning mechanism. In addition, we further discussed how to combine existing oversampling and ensemble concepts into the basic DELTA model in the pursuit of better

Fig. 4.10 tSNE
visualization of sample
distributions in original
sample space and DELTA
latent space. s is Calinski
and Harabaz score to show
the compactness. Data were
randomly downsampled to
5000 samples for
visualization

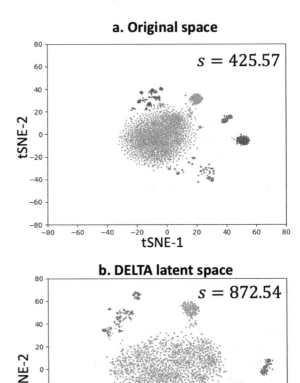

performances. The method proposed in this chapter provides new ideas and tools
for association inference and feature learning classification problems on imbalanced
datasets.

References

1. Korte A, Farlow A (2013) The advantages and limitations of trait analysis with gwas: a review.
   Plant Methods 9(1):1
2. Manolio TA, Collins FS, Cox NJ, Goldstein DB, Hindorff LA, Hunter DJ, McCarthy MI,
   Ramos EM, Cardon LR, Chakravarti A et al (2009) Finding the missing heritability of complex
   diseases. Nature 461 (7265):747–753

3. Ding X, Wang J, Zelikovsky A, Guo X, Xie M, Pan Y (2015) Searching high-order snp combinations for complex diseases based on energy distribution difference. IEEE/ACM Trans Comput Biol Bioinf (TCBB) 12(3):695–704

4. Goudey B, Rawlinson D, Wang Q, Shi F, Ferra H, Campbell RM, Stern L, Inouye MT, Ong CS, Kowalczyk A (2013) Gwis-model-free, fast and exhaustive search for epistatic interactions in case-control gwas. BMC Genomics 14(3):1

5. Guo X, Meng Y, Yu N, Pan Y (2014) Cloud computing for detecting high-order genome-wide epistatic interaction via dynamic clustering. BMC Bioinf 15(1):1

6. Bezzina CR, Barc J, Mizusawa Y, Remme CA, Gourraud JB, Simonet F, Verkerk AO, Schwartz PJ, Crotti L, Dagradi F et al (2013) Common variants at scn5a-scn10a and hey2 are associated with brugada syndrome, a rare disease with high risk of sudden cardiac death. Nature Genetics 45(9):1044–1049

7. He H, Garcia EA (2009a) Learning from imbalanced data. IEEE Trans Knowl Data Eng 21(9):1263–1284

8. Wang WYS, Barratt BJ, Clayton DG, Todd JA (2005) Genome-wide association studies: theoretical and practical concerns. Nat Rev Genet 6(2):109–118

9. McCarthy MI, Abecasis GR, Cardon LR, Goldstein DB, Little J, Ioannidis JPA, Hirschhorn JN (2008) Genome-wide association studies for complex traits: consensus, uncertainty and challenges. Nat Rev Genet 9(5):356–369

10. Lippert C, Listgarten J, Liu Y, Kadie CM, Davidson RI, Heckerman D (2011) Fast linear mixed models for genome-wide association studies. Nat Methods 8(10):833–835

11. Zhou X, Stephens M (2012) Genome-wide efficient mixed-model analysis for association studies. Nat Genet 44(7):821–824

12. Raskutti B, Kowalczyk A (2004) Extreme re-balancing for svms: a case study. ACM Sigkdd Explor Newslett 6(1):60–69

13. Tibshirani R (1996) Regression shrinkage and selection via the lasso. J Roy Stat Soc. Ser B (Methodol) 267–288

14. King G, Zeng L (2001) Logistic regression in rare events data. Polit Anal 9(2):137–163

15. Siva N (2008) 1000 genomes project. Nat Biotechnol 26(3):256–256

16. He H, Garcia EA (2009b) Learning from imbalanced data. IEEE Trans Knowl Data Eng 21(9):1263–1284. ISSN 1041-4347. https://doi.org/10.1109/TKDE.2008.239

17. Akata Z, Perronnin F, Harchaoui Z, Schmid C (2014) Good practice in large-scale learning for image classification. IEEE Trans Pattern Anal Mach Intell 36(3):507–520. ISSN 0162-8828. https://doi.org/10.1109/TPAMI.2013.146

18. Liu XY, Wu J, Zhou ZH (2009) Exploratory undersampling for class-imbalance learning. IEEE Trans Syst Man Cybern Part B (Cybern) 39(2):539–550, ISSN 1083-4419. https://doi.org/10.1109/TSMCB.2008.2007853

19. Chawla NV, Bowyer KW, Hall LO, Kegelmeyer WP (2002) Smote synthetic minority oversampling technique. J Artif Intell Res 16:321–357

20. Sun Y, Kamel MS, Wong AKC, Wang Y (2007) Cost-sensitive boosting for classification of imbalanced data. Pattern Recognit 40(12):3358–3378

21. Deng Y, Bao F, Kong Y, Ren Z, Dai Q (2017a) Deep direct reinforcement learning for financial signal representation and trading. IEEE Trans Neural Networks Learn Syst 28(3):653–664

22. Deng Y, Ren Z, Kong Y, Bao F, Dai Q (2017b) A hierarchical fused fuzzy deep neural network for data classification. IEEE Trans Fuzzy Syst 25(4):1006–1012

23. Bao F, Deng Y, Zhao Y, Suo J, Dai Q (2017a) Bosco: Boosting corrections for genome-wide association studies with imbalanced samples. IEEE Trans Nanobiosci 16(1):69–77. ISSN 1536-1241. https://doi.org/10.1109/TNB.2017.2660498

24. Galar M, Fernandez A, Barrenechea E, Bustince H, Herrera F (2012) A review on ensembles for the class imbalance problem: bagging-, boosting-, and hybrid-based approaches. IEEE Trans Syst Man Cybern Part C (Appl Rev) 42(4):463–484. ISSN 1094-6977. https://doi.org/10.1109/TSMCC.2011.2161285

25. Jo T, Japkowicz N (2004) Class imbalances versus small disjuncts. Acm Sigkdd Explor Newslett 6(1):40–49

26. Zhang J (1999) Adacost: misclassification cost-sensitive boosting. In: Proceedings of International Conference on Machine Learning. pp 97–105
27. Maloof MA (2003) Learning when data sets are imbalanced and when costs are unequal and unknown. In: ICML-2003 Workshop on Learning from Imbalanced Data Sets II
28. Ting KM (2002) An instance-weighting method to induce cost-sensitive trees. IEEE Trans Knowl Data Eng 14(3):659–665. ISSN 1041-4347. https://doi.org/10.1109/TKDE.2002.1000348
29. Zhou ZH, Liu XY (2006) Training cost-sensitive neural networks with methods addressing the class imbalance problem. IEEE Trans Knowl Data Eng 18(1):63–77. ISSN 1041-4347. https://doi.org/10.1109/TKDE.2006.17
30. Xu Ying L, Jian Xin W, Zhi Hua Z (2006) A cascade-based classification method for class-imbalanced data. J Nanjing Univ 42(2):148–155
31. Khan SH, Hayat M, Bennamoun M, Sohel FA, Togneri R (2018) Cost-sensitive learning of deep feature representations from imbalanced data. IEEE Trans Neural Networks Learn Syst 29(8):3573–3587
32. Huang C, Li Y, Loy CC, Tang X (2016) Learning deep representation for imbalanced classification. In: Proceedings of the IEEE Conference on Computer Vision and Pattern Recognition. pp 5375–5384
33. Gamazon ER, Wheeler HE, Shah KP, Mozaffari SV, Aquino-Michaels K, Carroll RJ, Eyler AE, Denny JC, Nicolae DL, Cox NJ et al (2015) A gene-based association method for mapping traits using reference transcriptome data. Nat Genetics 47(9):1091–1098
34. Bao F, Deng Y, Dai Q (2017b) Acid: association correction for imbalanced data in gwas. IEEE/ACM Trans Comput Biol Bioinf (99):1–1. ISSN 1545-5963. https://doi.org/10.1109/TCBB.2016.2608819
35. Deng Y, Bao F, Deng X, Wang R, Kong Y, Dai Q (2016) Deep and structured robust information theoretic learning for image analysis. IEEE Trans Image Proc 25(9):4209–4221
36. Deng Y, Dai Q, Liu R, Zhang Zengske, Hu S (2013) Low-rank structure learning via nonconvex heuristic recovery. IEEE Trans Neural Networks Learn Syst 24(3):383–396
37. Yang B, Fu X, Sidiropoulos ND, Hong M (2016) Towards k-means-friendly spaces: simultaneous deep learning and clustering. In: International Conference on machine learning
38. Kingma D, Ba J (2014) Adam: a method for stochastic optimization. *arXiv preprint* arXiv:1412.6980
39. Wu MC, Kraft P, Epstein MP, Taylor DM, Chanock SJ, Hunter DJ, Lin X (2010) Powerful snp-set analysis for case-control genome-wide association studies. Am J Hum Genet 86(6):929–942
40. Su Z, Marchini J, Donnelly P (2011) Hapgen2: simulation of multiple disease snps. Bioinformatics 27(16):2304–2305
41. Purcell S, Neale B, Todd-Brown K, Thomas L, Ferreira MAR, Bender D, Maller J, Sklar P, De Bakker PIW, Daly MJ et al (2007) Plink: a tool set for whole-genome association and population-based linkage analyses. Am J Hum Genet 81(3):559–575
42. Zhang Y (2012) A novel bayesian graphical model for genome-wide multi-snp association mapping. Genet Epidemiol 36(1):36–47
43. Siebert JP (1987) Vehicle recognition using rule based methods
44. Ramana BV, Babu MSP, Venkateswarlu NB (2012) A critical comparative study of liver patients from usa and india: an exploratory analysis. Int J Comput Sci Issues 9(2):506–516
45. Uci machine learning repository (2015) https://archive.ics.uci.edu/ml/datasets.html
46. Ganganwar V (2012) An overview of classification algorithms for imbalanced datasets. Int J Emerg Technol Adv Eng 2(4):42–47
47. Barua S, Islam MM, Yao X, Murase K (2014) Mwmote–majority weighted minority oversampling technique for imbalanced data set learning. IEEE Trans Knowl Data Eng 26(2):405–425
48. Criteo display ads ctr dataset (2015) https://www.kaggle.com/c/criteo-display-ad-challenge
49. Avazu mobile ads ctr dataset (2015) https://www.kaggle.com/c/avazu-ctr-prediction/data
50. Mailman MD, Feolo M, Jin Y, Kimura M, Tryka K, Bagoutdinov R, Hao L, Kiang A, Paschall J, Phan L et al (2007) The ncbi dbgap database of genotypes and phenotypes. Nat Genet 39(10):1181–1186

51. Abnet CC, Freedman ND, Hu N, Wang Z, Yu K, Shu XO, Yuan JM, Zheng W, Dawsey SM, Dong LM et al (2010) A shared susceptibility locus in plce1 at 10q23 for gastric adenocarcinoma and esophageal squamous cell carcinoma. Nat Genet 42(9):764–767

52. Luo D, Gao Y, Wang S, Wang M, Dongmei W, Wang W, Ming X, Zhou J, Gong W, Tan Y et al (2011) Genetic variation in plce1 is associated with gastric cancer survival in a chinese population. J Gastroenterol 46(11):1260–1266

53. Macosko EZ, Basu A, Satija R, Nemesh J, Shekhar K, Goldman M, Tirosh I, Bialas AR, Kamitaki N, Martersteck EM et al (2015) Highly parallel genome-wide expression profiling of individual cells using nanoliter droplets. Cell 161(5):1202–1214

54. Caliński T, Harabasz J (1974) A dendrite method for cluster analysis. Commun Statistics-theory Methods 3(1):1–27

55. Price AL, Zaitlen NA, Reich D, Patterson N (2010) New approaches to population stratification in genome-wide association studies. Nat Rev Genet 11(7):459–463

Chapter 5
Summary and Outlook

5.1 Summary of the Book

In this book, we disucssed three different situations of missing data, e.g. missing features, missing labels, and missing samples. For each missing situation, we proposed different computational recovery theories under the machine learning framework, to impute missing biological observations, improve data quality, and improve the interpretation of biological data. Specifically, major contributions are as follows:

- Biodata feature missings in life science research seriously limit the understanding of sample information. Especially in large-scale observation data, the problem of missing features will be more serious. Targeted at this challenge, this book introduces a deep recurrent autoencoder method, which achieves fast, low-memory cost, and high imputation accuracy on large-scale datasets, providing a new tool for large-scale data imputations. In addition to imputation, the model also has the functions of feature learning, correction of experimental batch effects, and data structure analysis, which provides a new choice for directly obtaining data features from degraded data. On 1.3 million mouse brain single-cell gene expression data, the proposed method fastly discovered all three major cell populations and 36 cell subtypes, providing a novel and effective method for analyzing the cell structure of the brain.
- Label missing and label polluting are unavoidable problems in life science research. Due to the restriction of the research problems, many label problems can only be alleviated through the recovery of calculation methods. In response to this problem, this book proposes a robust information theoretic learning method, which uses the information of existing labels to learn feature transformation functions and discriminative classifiers at the same time. At the same time, mutual information is used to constrain the data consistency between labels and features in hidden space to impute newly predicted labels. The proposed computational framework can be easily combined with the existing deep learning framework and structured sparse learning framework. On MRI image data of the brain with

© Tsinghua University Press 2021

F. Bao, *Computational Reconstruction of Missing Data in Biological Research*,
Springer Theses, https://doi.org/10.1007/978-981-16-3064-4_5

defaced labels, the proposed method robustly achieves better classification accuracy on brain regions.

- The lack of biological samples will severely restrict the understanding of data using general statistical methods, resulting in imbalanced data bias. In response to this problem, this book focuses on statistical association analysis and predictive analysis under missing sample data, and proposes structure-aware rebalancing. Under the statistical association analysis, this book proposes a feature structure-aware rebalancing method, which combines the downsampling and the statistical weighting strategies to perform unbiased correlation analysis under the weighted regression framework. In the imbalanced classification problem, this book proposes the sample structure-aware rebalancing, which combines deep feature transformation and clustering rebalancing strategies, and provides an end-to-end optimization method. The proposed method has achieved good performance in the identification of gastric cancer-assiciated genes and the classification of rare cells.

The methods proposed in this book are verified on both simulated data and real data, and are domenstrated how to apply to real collected biological data. It is hoped that these methods can provide new tools for life science research.

5.2 Future Works

With the rapid development in life science research, new types of observational data continue to appear, providing new opportunities for extensive understanding the operating principles of life. However, this also introduces new data analysis difficulties. Using computational methods to fill the gap between the shortcomings of observation technologies and biological discoveries has always been an important strategy for life data analysis. Under the new observation data types, computational methods will certainly continue to play an important role.

5.2.1 Computational Methods for Multi-modal Information Integration

Most of existing biological technologies only focus on characterizartion of life activities from one angle. Obviously, different observation methods can provide different perspectives of the same life process and provide complementary information. Different information integration is of great significance for a full-view understanding of life sciences. The bottleneck of traditional multiple observations is that they cannot provide different information for the same biological sample at the same time. Therefore, it is impossible to determine whether the information provided by different observation methods is caused by sample biases or true biological internal differences.

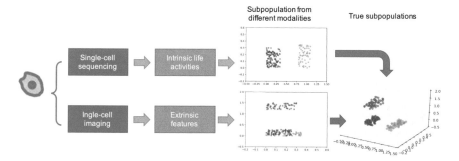

Fig. 5.1 An overview of multi-modal data integration

However, with the development of new life technologies, such as seqFISH, real-time multi-modality observations of the same sample have been possible. As shown in Fig. 5.1, through this method we can capture the morphological modality (images) and expression modality (genes) of cells at the same time, thereby achieving simultaneous analysis of the internal and external properties of the cells. It will greatly advance our understanding of cell functions.

5.2.2 Computational Methods to Understand Brain Functions

The nervous system is complex, and understanding how it works has been a long-standing challenge. In response to this problem, people have been trying to analyze it from different angles. Recently, a number of large projects focusing on neuroscience have been proposed, such as the American Brain Project, the European Brain Project, and the Chinese Brain Project, which provide important opportunities for further exploration of the functions of neuroscience. To analyze hundreds of millions of neurons in the brain, it will inevitably pose severe challenges to existing computational methods. Therefore, new approaches for neurological research problems will also be an important research direction.

Printed in the United States
by Baker & Taylor Publisher Services